PERMAFROST IN CANADA

CANADIAN BUILDING SERIES

Sponsored by the Division of Building Research
National Research Council, Canada
Editor: Robert F. Legget

PERMAFROST IN CANADA

Its Influence on
Northern Development

Roger J. E. Brown

UNIVERSITY OF TORONTO PRESS

© University of Toronto Press 1970
Printed in Canada
SBN 8020-1602-2

Preface

The author is a Research Officer in the Division of Building Research, National Research Council, Ottawa. He joined in 1953 as a member of the Division's Northern Research Group to study the distribution of permafrost in Canada and the physical factors affecting it. In the course of this work investigations have been carried out in many parts of northern Canada.

In September, 1956, he was granted two years' leave of absence by the National Research Council to pursue graduate studies toward obtaining a PH.D. degree in geography at Clark University in Worcester, Massachusetts. The first year of these studies was spent at the Graduate School of Geography, Clark University, taking formal course work for the degree. The second year was spent in library research on the PH.D. thesis at the Scott Polar Research Institute, Cambridge. The degree was conferred in 1961.

Because of the increasing interest in permafrost in Canada, Dr. Robert F. Legget, Director of the Division of Building Research, National Research Council until his retirement in 1969, suggested to the author the possibility of rewriting the thesis in book form. Through his efforts, the University of Toronto Press agreed to consider the manuscript for publication. This book, appearing eight years later in the Canadian Building Series, is the result of encouragement the author received from both Dr. Legget and the Press.

Since permafrost underlies one-half of Canada's land surface it is an important area of study for this country. This work attempts to survey the major scientific and engineering aspects of permafrost: its definition, origin, and occurrence, the climate and terrain of permafrost areas, economic and historical considerations, problems of building and transportation on permafrost, and its effects on mining and agriculture. Examples are given in these different spheres of activity to illustrate the major problems and various techniques employed in northern Canada to counteract the adverse effects of permafrost. It is not possible to give the whole story because of space limitations and gaps in the records, but the historical background and some details of the current situation are presented.

Before the Second World War there was little experience with permafrost in northern Canada as economic activities were underway at only a few widely scattered locations. The existence of permafrost suddenly became of major concern during the war in northern military operations; and interest increased steadily in the postwar period as the North opened up to new economic ventures, mostly in the Subarctic where permafrost is for the most part discontinuous but also in the Arctic where permafrost is continuous. These developments are progressing steadily and permafrost continues to be a vital factor on many fronts.

At the time of the writing of this book, activities in the Arctic are reaching new levels of intensity as tremendous reserves of oil are discovered and exploration continues for more. As economic developments grow in number and scope, it is becoming increasingly difficult to keep track of all the experiences with permafrost. It is hoped that errors of omission and fact will be forgiven since the author's intention was chiefly to convey an impression of the nature of permafrost and its impact on economic developments. This book is, however, only an introduction to the exhaustive documentation that awaits preparation to present the total picture of the effects of permafrost on the development of northern Canada.

Acknowledgments

The author is particularly indebted to Dr. Legget, for his valuable support and continued interest throughout graduate studies, the writing of the thesis, and subsequently this book. The two years' leave of absence granted by the National Research Council for the degree work, the second year being supported financially by the Council, was obtained largely through his efforts. His permission to use permafrost research material of the Division of Building Research and the facilities of this Division for the reproduction of all maps, diagrams, and photographs in this book is gratefully acknowledged. His assistance and inspiration played a large role in the ultimate publication of the book.

The author's association over the past sixteen years with his immediate professional colleagues, G. H. Johnston and the late J. A. Pihlainen, of the Northern Research Group, Division of Building Research, has also been of great value in the writing of this book. Much of the information was obtained by working with them both in the office and in the field. Mr. Pihlainen initiated and guided the permafrost investigations carried out by the Division in the early 1950s and was involved with permafrost until his death in 1964. The author profited greatly in working with him on various research projects. I am grateful to Mr. Johnston for reviewing the manuscript of this book and offering many constructive suggestions. Much of the current unreferenced information on the engineering aspects of permafrost in Northern Canada has been obtained in discussions and association with him. His knowledge and experience in this field is unequalled in Canada and this book could not have been written without his assistance.

The staff of the Graduate School of Geography at Clark University was of continued assistance during graduate studies and writing the thesis. The author wishes to record his special appreciation to the late Dr. R. J. Lougee who was staff adviser until his death in May, 1960. Dr. Lougee assisted in the selection of the topic and gave much valuable guidance in the writing and revision of the first draft of the text. After his passing, Dr. R. E. Murphy very kindly agreed to assume the position of adviser. His guidance in the final stages of completing the thesis is gratefully acknowledged. Finally, the author wishes to record his gratitude to Dr. S. VanValkenburg, former Chairman of the Graduate School of Geography, for his continued interest in the author's progress at Clark University and in the subsequent years.

The author is grateful to the Scott Polar Research Institute, Cambridge, for placing its facilities at his disposal in obtaining research material for the thesis. The Librarian, H. G. R. King, and his staff made available much published and unpublished material. Without this valuable and untiring support, the writing of the thesis would have been much more difficult. The author wishes to record his appreciation for the time spent by Dr. T. E. Armstrong, Senior Research Fellow, in helping to obtain and evaluate the Russian permafrost literature at the Institute.

In addition to the financial support granted by the National Research Council, the year of research at the Scott Polar Research Institute was supported also by predoctoral awards from the Royal Society of Canada and the Social Science Research Council of Canada. This generous assistance was indispensable to the successful completion of graduate studies.

During the writing of the book, many valuable suggestions and comments were offered by two gentlemen who have many years of experience in the North, G. W. Rowley, Department of Indian Affairs and Northern Development, Ottawa, and Professor K. B. Woods, Goss Professor of Engineering, School of Civil Engineering, Purdue University, Lafayette, Indiana. Both spent considerable time reading the thesis manuscript and suggesting revisions for the book. Their assistance is gratefully acknowledged.

The author appreciates the continued interest of Miss Francess G. Halpenny and the expert assistance of Miss Margaret Gillies and Mrs. Diana Swift Sewell of the University of Toronto Press. He is also grateful to Mr. F. Crupi, Division of Building Research, National Research Council for drafting the maps and diagrams.

Finally, the author acknowledges the valuable assistance and support given by his wife through the entire period of studies leading to the degree, and later in the writing of the book. She provided continued encouragement as well as spending many hours in assisting with the revision, editing and proofreading of the manuscript.

Contents

PERMAFROST IN CANADA

1 Nature of Permafrost

About one-half of Canada lies in the permafrost region. This area (whose most prominent natural features, political divisions, and settlements, are shown in Figures 1 A, B, C, D) comprises the Yukon and Northwest Territories and the northern parts of most of the provinces, with a sparse population scattered over vast distances. Economic development and other human activities have been hampered by a number of factors, chiefly severity of climate, remoteness from the populated south, inaccessibility, and permafrost.

Since the end of the Second World War, and particularly during the past decade, Canada has become increasingly aware of her northern frontier. Despite more intensive activity, however, it is unlikely that this region will be developed to the same degree as southern Canada, although the wealth of natural resources, particularly minerals, will encourage such development.

The existence of permafrost usually necessitates modifications in conducting various tasks described in the following chapters because of the need to counteract its adverse effects. Thus, costs are higher than in temperature regions where there is no permafrost. Increasing knowledge of permafrost properties and characteristics has raised the level of technology to the point where it is possible to build virtually any structure in the permafrost region. Even though new methods will probably bring the cost of many operations nearer to those in temperate regions, permafrost will continue to be an adverse factor in the development of northern Canada.

A DEFINITION AND ORIGIN

The term "permafrost" was coined by S. W. Muller, who used it as a convenient short form of "permanently frozen ground" (Muller 1945). It is a term used to describe the thermal condition of earth materials, such as soil and rock, when their temperature remains below 32° F continuously for a number of years (Muller 1945; Pihlainen and Johnston 1963; Shvetsov 1959; Stearns 1966; Sumgin et al. 1940). Ice is an important component of permafrost but all water in the ground

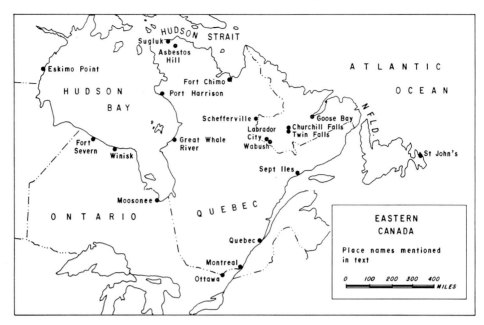

FIGURE 1A Eastern Canada.

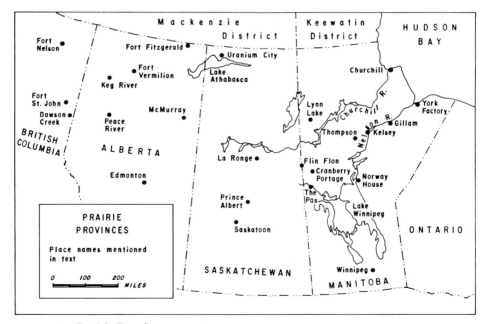

FIGURE 1B Prairie Provinces.

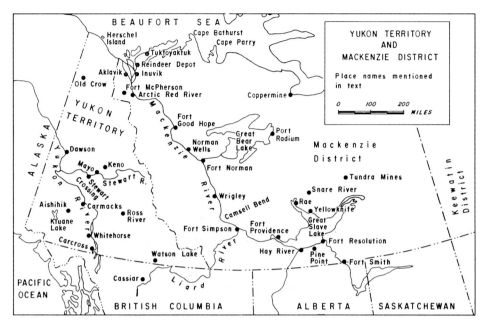

FIGURE 1C Yukon Territory and Mackenzie District.

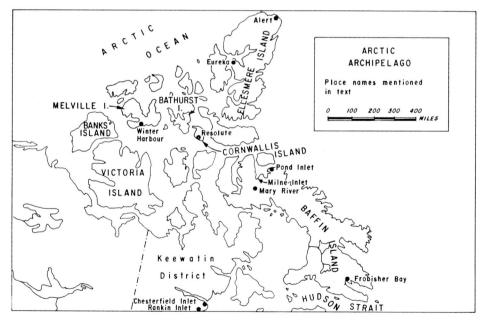

FIGURE 1D Arctic Archipelago.

does not freeze at 32° F if it contains impurities or is under pressure. Although freezing implies a change of state, the presence of unfrozen water in the soil at temperatures below 32° F causes semantic difficulties in the use of the term. Despite these problems, it is generally agreed that defining permafrost solely on the basis of temperature is the most workable arrangement.

Permafrost includes ground that freezes in one winter, and remains frozen through the following summer and into the next winter. This is the minimum limit for the duration of permafrost; it may be only a few inches thick. At the other end of the scale, permafrost is thousands of years old and hundreds of feet thick. The mode of formation of such old and thick permafrost is identical to that of permafrost only one year old and a few inches thick. In the case of the former, even a small negative heat imbalance at the ground surface each year results in a thin layer being added annually to the permafrost. After several thousands of years have elapsed, this process repeated annually can produce a layer of permafrost hundreds of feet thick. However, the permafrost does not increase in thickness indefinitely. Rather a quasi-equilibrium is reached whereby the downward penetration of frozen ground is balanced by heat from the earth's interior. Above the permafrost is a surface layer of soil or rock, called the "active layer," shown in Figure 2, which thaws in summer and freezes in winter. Its thickness depends on the same climatic and terrain features that affect the permafrost.

In recent years, it has been realized that permafrost is not necessarily permanent. Changes in climate and terrain can cause the permafrost to thaw and disappear. Thus the term "perennially frozen ground" is now used instead of "permanently frozen ground." This change in terminology is also evident in Russian literature where the term *mnogoletnemërzlyy grunt* meaning "perennially frozen ground" is replacing *vechnaya merzlota* meaning "eternal frost."

The origin of permafrost is not well understood but it is suspected that it first appeared during the cold period at the beginning of the Pleistocene. During subsequent periods of climatic fluctuations, corresponding changes have occurred in its extent and thickness. In the north, where winters are long and severe and summers are short, the depth of winter frost penetration will be generally greater than the summer thawing of the ground. In this regime, a layer of frozen ground will persist through the summer and another layer will be added to it during the following winter. If the climate remains sufficiently cold over a period of years a considerable thickness of perennially frozen ground will accumulate.

For example, it can be calculated, on the basis of simple conduction theory, that at Resolute, NWT, on Cornwallis Island, about 10,000 years was required for the ground to freeze to a depth of 1,300 feet, the present estimated thickness of the permafrost. A more precise estimate would have to take account of such factors as the latent heat of fusion of ice, proximity of the ocean, fluctuations in mean annual air temperature since the initiation of permafrost accumulation, and changes in terrain conditions (Brown and Johnston 1964). If the climate becomes warmer, some or all of this permafrost may disappear, but if the climate becomes even colder, the permafrost may increase in thickness at a faster rate. At present it is known to be diminishing in some areas and increasing in others.

B   DISTRIBUTION AND OCCURRENCE

1 *Geographic Extent and Thermal Regime*

Permafrost underlies 20 per cent of the world's land area, being widespread in North America, Eurasia, and Antarctica. In the northern hemisphere it occurs mostly in Canada and the Soviet Union, each country having about one-half of its total land area underlain by it (R. J. E. Brown 1967*b*, Permafrost Map shown in Figure 3). It is found also in most of Alaska, Greenland, northern Scandinavia, and in Outer Mongolia and Manchuria. It occurs also at high elevation in mountainous regions in other parts of the world.

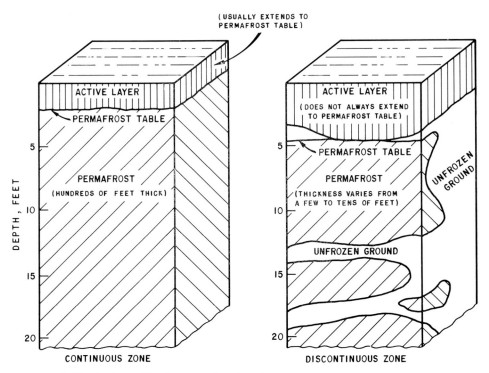

FIGURE 2 Typical profiles in permafrost region.

The permafrost region is divided into two zones – discontinuous in the south, and continuous in the north. Various criteria have been used in North America to delineate the division between these zones. The method employed in Canada by R. J. E. Brown 1967*b*, Permafrost Map (see Figure 3) is based on the arbitrary selection by Russian permafrost investigators of the minus 5° C (23° F) isotherm of mean annual ground temperature measured just below the zone of annual variation (Bondarev 1959). This criterion appears to apply to the Canadian situation according to the presently known distribution of permafrost but more field observations are required to assess the validity of its uses in this country. American investigators have used the Russian and other criteria to delineate the permafrost zones in Alaska (Black 1954, Ferrians 1965).

In the continuous zone, permafrost occurs everywhere beneath the ground surface except in newly deposited unconsolidated sediments where the climate has just begun to impose its influence on the ground thermal regime. The thickness of permafrost is about 200 feet at the southern limit of the continuous zone increasing steadily to more than 1,000 feet in the northern part of the zone as Figure 4 shows. The active layer generally varies in thickness from about 1½ to 3 feet and usually extends to the permafrost table. The temperature of the permafrost in this zone, at the depth at which annual fluctuations become virtually imperceptible (i.e., less than about 0.1° F) known as the level of zero annual amplitude, ranges from about 23° F in the south to about 5° F in the extreme north.

In the discontinuous zone, frozen and unfrozen layers occur together (see Figure 4). In the southern fringe of this zone, permafrost occurs in scattered

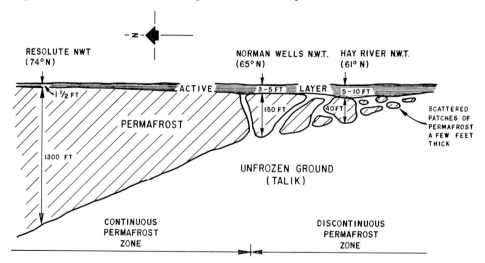

FIGURE 4 Typical vertical distribution and thickness of permafrost.

LEGEND

PERMAFROST

CONTINUOUS PERMAFROST ZONE

SOUTHERN LIMIT OF CONTINUOUS PERMAFROST ZONE

DISCONTINUOUS PERMAFROST ZONE
WIDESPREAD PERMAFROST

SOUTHERN FRINGE OF PERMAFROST REGION

SOUTHERN LIMIT OF PERMAFROST

PATCHES OF PERMAFROST OBSERVED IN PEAT
BOGS SOUTH OF PERMAFROST LIMIT

PERMAFROST AREAS AT HIGH ALTITUDE IN CORDILLERA
SOUTH OF PERMAFROST LIMIT

CLIMATE

MEAN ANNUAL AIR TEMPERATURE, °F

PHYSIOGRAPHIC REGIONS

BOUNDARY OF REGIONS

(1) PRECAMBRIAN SHIELD

(2) HUDSON BAY LOWLAND

(3) INTERIOR PLAINS

(4) CORDILLERA

(5) ARCTIC ARCHIPELAGO

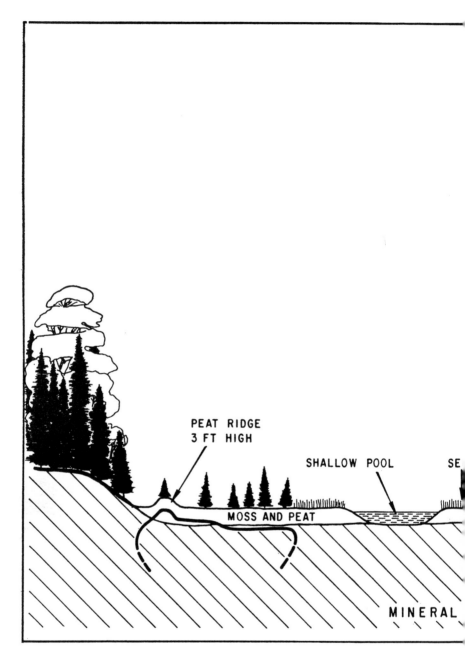

PEAT RIDGE
3 FT HIGH

SHALLOW POOL

SE

MOSS AND PEAT

MINERAL

FIGURE 5 Profile through typical peatland in southern fringe of discontinuous zo

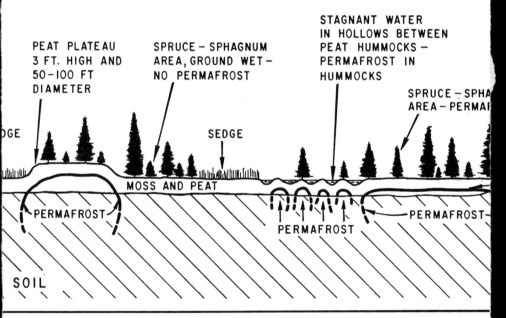

PEAT PLATEAU
3 FT. HIGH AND
50-100 FT
DIAMETER

SPRUCE – SPHAGNUM
AREA, GROUND WET –
NO PERMAFROST

STAGNANT WATER
IN HOLLOWS BETWEEN
PEAT HUMMOCKS –
PERMAFROST IN
HUMMOCKS

SPRUCE – SPHA
AREA – PERMA

)GE

SEDGE

MOSS AND PEAT

PERMAFROST

PERMAFROST

PERMAFROST

SOIL

ne showing interaction of permafrost and terrain factors.

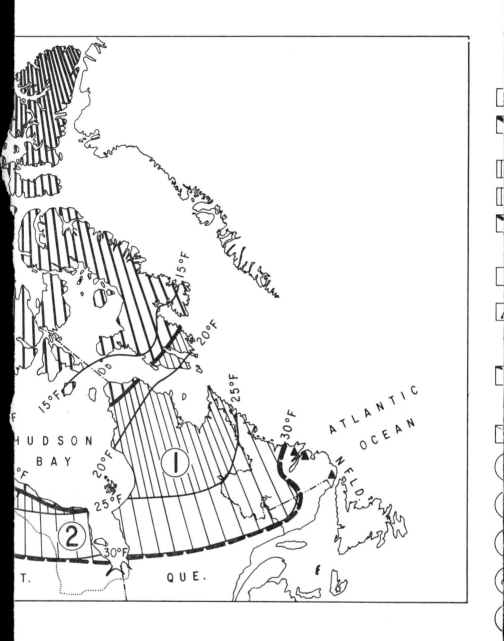

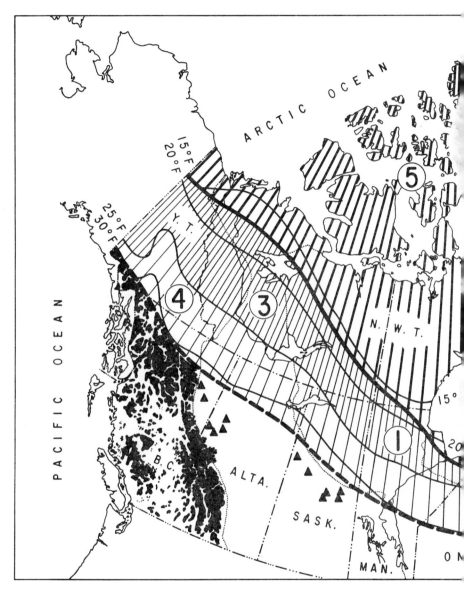

FIGURE 3 Permafrost in Canada.

SPRUCE, POPLAR,
JACKPINE AND BIRCH

GNUM
ROST

DEPTH TO PERMAFROST
TABLE IS $1\frac{1}{2}$ – 3 FT .
PERMAFROST LAYER VARIES
IN THICKNESS FROM A FEW
INCHES TO SEVERAL FEET

islands a few square feet to several acres in size and is confined to certain types of terrain, mainly peatlands (see Figure 5). Other occurrences are associated either with the north-facing slopes of east-west oriented valleys, or with isolated patches in forested stream banks, apparently in combination with increased shading from summer thawing and reduced snow cover. Northward it becomes increasingly widespread in a greater variety of terrain types. Permafrost varies in thickness from a few inches or feet at the southern limit to about 200 feet at the boundary of the continuous zone, with unfrozen layers sometimes occurring between layers of permafrost. The depth to the permafrost table ranges from about two feet to ten feet or more depending on local climatic and surface terrain conditions. The active layer does not always extend to the permafrost table. The temperature of the permafrost in the discontinuous zone at the level of zero annual amplitude generally ranges from a few tenths of a degree below 32° F at the southern limit to 23° F at the boundary of the continuous zone.

The thickness and temperatures of permafrost at various locations in Canada are listed in Table 1. The thickest permafrost recorded in Canada occurs at Winter Harbour on Melville Island, NWT. The thickest known permafrost in the northern hemisphere is in eastern Siberia about 550 miles northwest of Yakutsk where it has been recorded to a depth of 5,000 feet (Yefimov and Dukhin 1966).

In the continuous permafrost zone the mean annual ground temperature decreases steadily from the ground surface to a depth of from 50 to 100 feet as can be seen in Figure 6. Below this depth the temperature increases steadily under the influence of the heat from the earth's interior. After a time lag determined by depth and local terrain conditions, fluctuations in air temperature during the year produce corresponding temperature fluctuations about the mean annual ground temperature to depths of some 20 to 50 feet. The amplitude of these fluctuations decreases with depth to less than 0.1° F at the level of zero annual amplitude. Below this, ground temperatures change only in response to long term climatic changes extending over centuries. In the discontinuous zone the situation differs from the continuous zone because the permafrost is interspersed with areas and layers of unfrozen ground, and it is thinner varying in depth from only a few feet to a maximum of about 200 feet. Observations in this zone, such as at Thompson and Kelsey in northern Manitoba, indicate that permafrost temperatures are in many cases virtually isothermal and only a few tenths of a degree to one degree or so below 32° F.

The most southerly extent of permafrost in Canada, excluding the Western Cordillera, is about latitude 51° N to 52° N around James Bay. West of Hudson Bay the southern limit extends northwest through the northern parts of the Prairie Provinces and British Columbia and the southwest corner of Yukon Territory.

TABLE 1
Thickness and Temperature of Permafrost

Location	Thickness of permafrost (feet)	Ground temperature (°F) at various depths (feet)	Mean annual air temperature (°F)
1. Aishihik, YT	50–100	28.3 (20)	24.5
2. Asbestos Hill, PQ	>900	19–20 (50–200)	17
3. Churchill, Man.	100–200	27.5–28.9 (25–54)	19
4. Dawson YT	200	—	23.6
5. Fort Simpson, NWT	40	35.4–33.2 (0–5)	25.0
6. Fort Smith, NWT	unknown	about 32 (15)	26.2
7. Fort Vermilion, Alta.	nil	39.8–38.9 (0–5)	28.2
8. Inuvik, NWT	>300	26 (25–100)	15.6 (Aklavik)
9. Keg River, Alta.	5	31–32 (5)	31
10. Kelsey, Man.	50	30.5–31.5 (30)	25.5
11. Mackenzie Delta, NWT	300	23.8–26.5 (0–100)	15.6 (Aklavik)
12. Mary River, NWT– Baffin Island	unknown	10 (30)	6.3 (Pond Inlet)
13. Milne Inlet, NWT– Baffin Island	unknown	10 (50)	6.3 (Pond Inlet)
14. Norman Wells, NWT	150–200	26–28.5 (50–100)	20.8
15. Port Radium, NWT	350	—	19.2
16. Rankin Inlet, NWT	1000	15–17 (100)	11.2 (Chesterfield Inlet)
17. Resolute, NWT– Cornwallis Island	1300	10–8.5 (50–100)	2.8
18. Schefferville, PQ	>250	30–31.5 (25–190)	23.9
19. Thompson, Man.	50	31–32 (25)	24.9
20. Tundra Mines Ltd., NWT	900	29 (325)	17
21. Uranium City, Sask.	30	31–32 (30)	24
22. United Keno Hill Mines Ltd., YT	450	28–29.3 (100–200)	24.2 (Elsa)
23. Winter Harbour, NWT– Viscount Melville Island	1500	—	—
24. Yellowknife, NWT	200–300	31.4 (40)	22.2

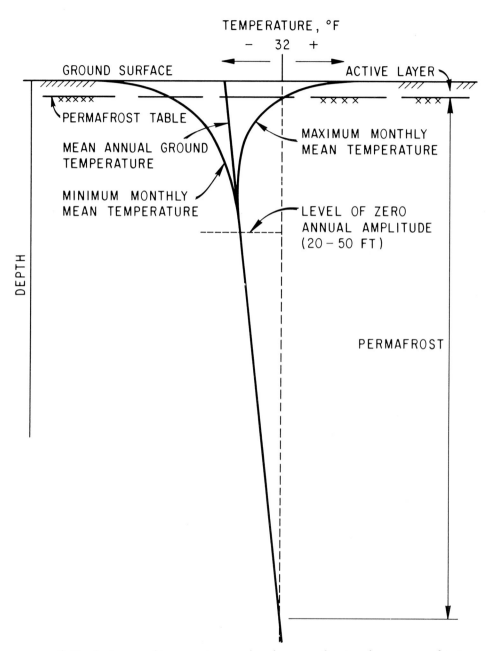

FIGURE 6 Typical ground temperature regime in permafrost and non-permafrost areas.

It is postulated that the division between the discontinuous and continuous permafrost zones extends across the northwest peninsula of Labrador-Ungava and the southeast tip of Baffin Island but this has not been verified by field observations (Brown 1967). West of Hudson Bay it extends from northern Ontario in a northwesterly direction to Great Bear Lake and on westward across the Mackenzie River and northern Yukon Territory.

In the Cordillera, the distribution of permafrost varies with altitude as well as latitude. On the map (Figure 3) the southern limit of permafrost marks approximately the boundary of permafrost occurrence at valley bottom levels. South of this line permafrost is not present in lowlands and valley bottoms but exists at higher altitude. The lower altitudinal limit of permafrost rises progressively from north to south. Throughout northern British Columbia, south of the southern limit of permafrost on the map, field observations indicate that the lower altitudinal limit of permafrost is uniformly at about 4,000 feet above sea level. Below this elevation, scattered permafrost islands occur only in specific types of terrain. In the southern part of the Cordillera the lower altitudinal limit of permafrost has been estimated to increase steadily from an elevation of 4,000 feet at latitude 54°30′ N to about 7,000 feet at the forty-ninth parallel.

With increasing elevation the distribution of permafrost changes progressively from scattered islands to widespread and, finally, continuous. In mountainous regions there is the added problem of the differences between slopes of different orientation. Permafrost is more widespread and thicker on north-facing slopes than on south-facing slopes. Snow cover, which has considerable influence on permafrost, is heavier on windward than leeward slopes. Other terrain factors also vary from one slope to another to complicate the distribution of permafrost.

It is believed that permafrost occurs at the summit of Mont Jacques-Cartier (4,160 feet) in the Gaspé Peninsula, PQ. This is the highest elevation south of the permafrost limit in eastern Canada. It is suspected of being the only location east of Hudson Bay which is of sufficiently high altitude to support the existence of permafrost at its summit. (Permafrost is reported to exist at the summit of Mount Washington in New Hampshire – 6,288 feet above sea level.)

At the southern limit of permafrost in Canada the known occurrences seem to be in reasonable equilibrium with the present environment. To date no occurrences which represent radically different conditions have been described south of the permafrost region. There are a few unsubstantiated reports, however, of isolated bodies of permafrost lying at depth beneath the ground surface. If such permafrost does exist, it formed probably during a previous period of cooler climate and lies at a depth below that affected by the current climate. No evidence of these bodies of permafrost exists on the ground surface and they would be detected only by mining operations or ground temperature measurements.

2 *Surface Features*

Many distinctive features occur on the ground surface in the permafrost region which are indicative of underlying conditions. Investigators in North America and Eurasia have written many accounts of these forms which are described in this brief summary. The reader is directed to the extensive literature and to the review by Stearns (1966) from which much of the information in this account is derived. Only a few references are cited here, including some notable Canadian contributions.

A comprehensive classification of patterned ground and a review of suggested origins was presented by Washburn (1956). He considered patterned ground as a collective term to include polygons, circles, nets, steps, stripes, and mounds. These features are not restricted to permafrost regions but they attain their best development here.

Polygons occur in many forms and sizes furnishing a pattern that is widespread throughout the cold regions. Washburn classified them as either raised-centre (borders are cracks or fissures) or depressed-centre (raised borders) polygons, in which the surface materials may be sorted or non-sorted. They may be of all sizes up to 100 feet in diameter. In some places primary polygonal networks are divided by secondary, internal polygons. The existence of permafrost is not essential for their formation but the largest ones are found in the permafrost region. They are particularly well developed in the continuous zone where ice wedge tundra polygons, both raised-centre and depressed-centre, are widespread (see Figure 7).

Circles may be sorted or non-sorted. This patterned ground has a mesh with two- to twelve-foot diameters and sometimes larger circular elements. The sorted circles or stone rings have clean stone borders surrounding fine-grained soil sometimes with gravel or pebbles. Non-sorted circles also have a circular mesh but the stone border is absent. They include stony earth circles with centres of silt and some sand and gravel which form from plugs of soil in the active layer. Both sorted and non-sorted circles form above permafrost, but permafrost is not essential.

Steps and stripes vary in width from a few inches to several feet, occur on moderate to steep slopes and may be sorted or non-sorted. The steps are parallel to the contour while the stripes extend downslope. The non-sorted steps consist of an earth core covered by moss and other plants, and the difference in elevation forms the steps. Non-sorted stripes are alternating parallel lines of vegetation and bare ground. Solifluction stripes, lobes, and scars can be noted where soil is moving slowly downslope under the combined forces of frost action, gravity, and fluid flow. Frost boils are also a common pattern caused by intense frost action in the active layer which brings fine-grained soil to the surface in

concentrated circular deposits. Mounds caused by intense frost heaving in the
active layer also form microrelief of three to four feet.

The most spectacular landforms associated with permafrost are pingos. These
conical hills, which may grow through several centuries or thousands of years
to more than 100 feet in height and one-quarter mile in diameter, are found in
the Mackenzie Delta region, where more than 1000 have been mapped (Stager
1956), Northern Alaska, Northeast Greenland, and Siberia. They have ice centres
covered with soil and vegetation; they are often cracked diametrically at the top
and they become widely cratered later in their lifespan. Two types of pingos classified according to origin have been described. Closed system pingos occur in the
continuous zone in flat, poorly drained terrain such as old lake bottoms or river
deltas where water in the underlying thaw basins is forced toward the surface
by the surrounding aggrading permafrost (Mackay 1962) (see Figure 8). Open
system pingos form in the discontinuous zone where groundwater in unfrozen
zones between permafrost layers is forced to the surface and an ice mass accumulates (Müller 1959).

FIGURE 7 Aerial view from about 1,000 feet of tundra ice-wedge polygons fifty to
a hundred feet in diameter in continuous permafrost zone near Tuktoyaktuk, NWT.
The cracks or fissures are underlain by wedge-shaped masses of ice up to three feet
wide, extending to depths of ten to twenty feet.

Surface features associated with permafrost in the discontinuous zone are less numerous and distinctive than in the continuous zone. Low hills or knolls of perennially frozen peat, about ten feet or less in height, designated by the Swedish term "palsa," occur as permafrost islands in peatlands and peat bogs. They form by the combined action of peat accumulation and ice segregation in the underlying mineral soil (Sjörs 1959a, b and 1961) (see Figure 9). These mounds grow and coalesce to form ridges and plateaus which form distinctive ground surface patterns particularly widespread in the northern part of the discontinuous zone (R. J. E. Brown 1968a) (see Figures 10, 11, 12).

The thawing of permafrost also forms distinctive landforms both in the discontinuous and continuous zones. "Thermokarst" is the term used for uneven land subsidence caused by the melting of ground ice. The resulting ground surface is very hummocky and uneven resembling the karst topography found in limestone areas where sink holes provide a landscape of many small lakes. Beaded streams are particularly prevalent in the continuous zone where blocks of massive ground ice melt in the stream bed forming enlargements along its course (Figure 13). The thawing of ice wedges of polygonal ground gives a distinctive pattern of regular

FIGURE 8 Aerial view from about 800 feet of a pingo around Tuktoyaktuk near the arctic coast, east of the Mackenzie River delta, in continuous permafrost zone. This pingo is about 130 feet high and has a crater on the summit. It is located in a lake bed and has a core of massive ice.

hillocks termed "baydzherakhi" (cemetery mounds) in the USSR. Massive slumps may occur on the hillsides where large masses of ground ice are exposed to melting by the disturbance of the surface vegetation. Disturbance of the surface vegetation by man in areas of ground ice concentration can also cause slumping as shown in Figure 14.

3 *Ground Ice*

Ice is a very important component of permafrost occurring in many forms (Pihlainen and Johnston 1963). It may take the form of layers or lenses ranging from hairline breadth, scarcely visible to the naked eye, to several feet in thickness (as shown in Figure 15). The layers are mostly horizontal but diagonal and vertical layers can also occur. These ice formations are normally observed in fine-grained soils. Extremely high ice contents are prevalent in these soils but may be difficult to discern. In some cases, such as with silty soils the volume of ice may be as much as six times that of the soil (Pihlainen and Johnston 1954). Other forms are as coatings or films on small particles, stones or boulders, or as indi-

FIGURE 9 Mature palsa (peat mound) in the southern fringe of discontinuous permafrost zone near Great Whale River, PQ. The peat is three feet thick overlying silty clay soil. The depth to permafrost in the palsa is about 1½ feet. The permafrost in the palsa is probably between ten and twenty feet thick. No permafrost occurs in the surrounding terrain.

vidual ice crystals or inclusions in the cavities in the soil. Such occurrences are usually found in coarse-grained deposits.

Some of the more spectacular ice deposits are found as random or irregularly oriented layers tens of feet thick, vertical wedges, and variable blocks or chunks sometimes ten to hundreds of feet in horizontal and vertical extent. The surface features, such as ice wedge polygons and pingos, associated with these ice forms have already been described. Numerous studies have been made on these various types of massive ice in North America (Lachenbruch 1966, Mackay 1966, Péwé 1966) and in Siberia (Dostovalov and Popov 1966). Russian investigators have undertaken a comprehensive classification of ground ice with three main categories and many subdivisions (Shumskiy and Vtyurin 1966).

The normally rock-like qualities and relatively high strength exhibited by permafrost can be attributed in part to the cementing action of the ice which binds the soil particles into a solid mass, as well as to composition, texture, and temperature. The mechanical properties of frozen ground, in which ice fills some

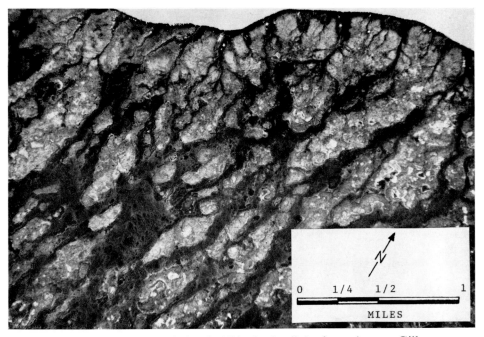

FIGURE 10 Aerial photograph (scale 1⅞ inch: 1 mile) of terrain near Gillam, Man., on Nelson River in discontinuous permafrost zone. Light-toned areas are forested peat plateaus with permafrost about eighty feet thick; the active layer is about two feet thick. Dark areas are wet, sedge-covered depressions with no permafrost.

FIGURE 11 Aerial view from about 400 feet of the terrain shown in Figure 10. Forested peat plateau with permafrost at top of photograph. Wet, sedge-covered depression with no permafrost in middle of photograph.

FIGURE 12 Ground view of terrain shown in Figure 10. Man is standing on peat plateau with permafrost. Wet, sedge-covered depression with no permafrost in foreground.

or all of the interstitial space between soil grains, therefore, tend to approach those of ice. The strength of frozen ground increases with decrease in temperature and, in general, with increase in moisture (ice) content. For some soils, such as clays, the increase in strength is relatively small at temperature just below freezing, mainly because of the amount of unfrozen water in the material. Frozen sands that are well cemented by ice usually have considerably greater strength than fine-grained materials, particularly at temperatures near thawing (Vyalov 1959).

C   CLIMATE AND TERRAIN

Permafrost exists as a result of a thermal condition which is reflected in a mean ground temperature that never rises above 32° F. Because temperature is an index of the level of heat storage, it fluctuates in response to the heat losses and heat gains at the ground surface. These follow a cyclical pattern on a daily as well as a yearly basis. This heat exchange at the ground surface is highly variable in respect of climatic factors and is also affected by terrain or surface conditions including the

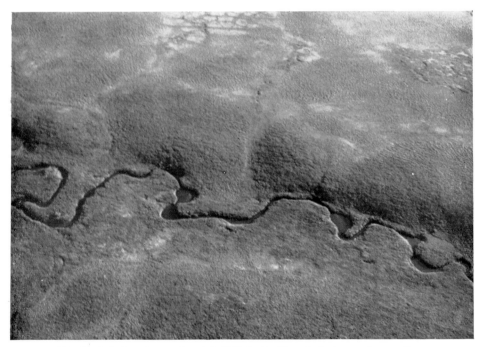

FIGURE 13 Aerial view from about 1,000 feet of beaded stream in continuous permafrost zone near Tuktoyaktuk. Enlargements or "beads" in stream are caused by melting of blocks of ground ice.

thermal properties of the ground itself, often referred to as the geothermal conditions. It is highly variable not only with time at any one point but may also be variable from point to point on the ground (Legget *et al.* 1961).

1 *Climate*

Climate is a basic factor in the formation and existence of permafrost. Observations indicate a broad relation between mean annual air and ground temperatures in permafrost regions. Snow cover is prominent along with the complex energy exchange regime at the ground surface causing the mean annual ground temperature, measured at the level of zero annual amplitude, to be several degrees warmer than the mean annual air temperature. Local microclimates and terrain conditions cause variations ranging from about 2 to 10 degrees, but a value of 6° F can be used as an average figure (see Table 1). For example, the mean annual air temperature at Resolute, NWT, in the continuous permafrost zone is 2.8° F and the mean annual ground temperature at the 100 foot depth is 8.5° F. Thompson, Manitoba, in the discontinuous zone, has a mean annual air temperature of 24.9° F and a mean annual ground temperature in the permafrost of about 31° F.

FIGURE 14 Slump in gravel pit at Inuvik, NWT. Removal of the natural vegetation has caused the underlying massive ground ice in the permafrost to melt and a depression to form. Note coat and hat for scale.

Present knowledge of the southern limit of permafrost indicates that it coincides roughly with the 30° F mean annual air isotherm. West of Hudson Bay, the southern limit on the map (Figure 3) is based on a considerable number of field observations. East of Hudson Bay, there are few field observations and the southern limit is shown to coincide with the 30° F mean annual air isotherm. Southward, permafrost occurrences are rare and small in size because the climate is too warm. Between the 30° F and 25° F mean annual air isotherms, permafrost is restricted mainly to the drier portions of peatlands and peat bogs because of the special insulating properties of peat (see Figure 5). Scattered bodies of permafrost also occur on some north-facing slopes and in some heavily shaded areas.

In the vicinity of the 25° F mean annual air isotherm the difference of 6° F between the mean annual air and ground temperature produces a mean annual ground temperature of a fraction of a degree below 32° F in most types of terrain. From the 25° F mean annual air isotherm, northward to the continuous zone, permafrost becomes increasingly widespread and thicker, and the mean annual ground temperature decreases.

FIGURE 15 Massive horizontal layers of ground ice in perennially frozen soil face exposed by landslide. The ice is protected from melting by overhanging vegetation. This exposure is located in a soil slump at the base of the Richardson Mountains which flank the west side of the Mackenzie River delta about fifteen miles southwest of Aklavik, NWT.

There is virtually no field information on the boundary separating the discontinuous and continuous zones. The line on Figure 3 is located at the approximate location of the 17° F mean annual air isotherm to correspond with a mean annual ground temperature of 23° F. Field observations along the coast of Hudson Bay in Ontario and Manitoba indicate a narrow band of continuous permafrost south of the 17° F mean annual air isotherm. From this isotherm northward, permafrost is continuous and increasingly thicker, and the mean annual ground temperature decreases.

Over a long period of time, a change in the climate represented by a change in the mean annual air temperature can result in a significant change in the extent and thickness of permafrost. Geothermal gradients ranging from about 1° F/40 feet to 1° F/300 feet – depending to some degree on the type of soil or rock – have been observed in permafrost regions in the northern hemisphere. A change of 1° F in the mean annual air temperature, for example, would result, over a long period of time, in a change of 1° F in the mean annual ground temperature. This would cause a change in permafrost thickness of approximately 40 to 300 feet.

Variations in cloud cover throughout the permafrost region may cause significant differences in the amount of solar radiation received by the ground surface. Cloud cover is generally greater east of Hudson Bay than in western Canada. This could influence the distribution of permafrost but no detailed information is available.

Microclimatic factors are also important in influencing the distribution of permafrost. Net radiation, evaporation (including evapotranspiration) – condensation, and conduction-convection are also elements of the energy exchange regime at the ground surface (Brown 1965). Although they are climatic in origin, their contribution to the ground thermal regime is determined by the nature of the ground surface and thus can be considered as terrain factors. Precipitation is another factor influencing permafrost. It is also treated as a terrain factor, rain under drainage and snow as a layer lying on the ground surface.

2 Terrain

The broad pattern of permafrost distribution is determined by climate but terrain conditions are responsible for local variations. In the discontinuous zone variations in terrain conditions are responsible for patchy occurrence of permafrost, size of permafrost islands, depth to the permafrost table, and thickness of permafrost (Legget et al. 1961). In the continuous zone, the thermal properties of the peat and other terrain factors assume a relatively minor role and the thermal properties of the ground as a whole, together with the climate, become dominant. Terrain factors that affect permafrost conditions include relief, vegetation, drainage, snow cover, fire, soil and rock type, and glacier ice. In its turn, permafrost influences the terrain.

a / Relief
Relief influences the amount of solar radiation received by the ground surface and the accumulation of snow. The influence of orientation and degree of slope is particularly evident in mountainous regions. In the discontinuous zone this may result in permafrost occurring on north-facing slopes but not on adjacent slopes facing south. In the continuous zone permafrost is thicker and the active layer thinner on north-facing slopes. Variations in snowfall and accumulation on the ground may modify these patterns. Smaller scale variations in relief cause similar situations elsewhere in the permafrost region. Permafrost may occur for example, in the north-facing bank of even a small stream but not in the opposite south-facing bank. Similar differences can occur even in areas of intensive micro-relief such as peat mounds and peat plateaus.

    Permafrost affects the relief of ground surface configuration. Solifluction and other downslope mass movements of earth material over permafrost surfaces alter the ground surface configuration. Thermokarst is frequently an active process in areas where large masses of ground ice exist. The melting of this ice and result-ing subsidence of the ground produce undulations and hollows which considerably alter the surface configuration.

b / Vegetation
Vegetation affects permafrost in various ways and is one of the more obvious indi-cators of subsurface conditions (R. J. E. Brown 1966a, Tyrtikov 1959). The most obvious effect of vegetation is its role of shielding the permafrost from solar heat. This protection is provided mainly by the insulating properties of the wide-spread moss and peat cover. Removal or even disturbance of this surface cover causes degradation of the underlying permafrost. In the discontinuous zone this may result in the disappearance of bodies of permafrost. In the continuous zone the permafrost table will be lowered. The predominance of moss and peat in pro-tecting the permafrost from atmospheric heat is demonstrated by the fact that little change occurs in the depth of the permafrost table when trees and brush are removed provided that the moss and peat are not disturbed. A fire may burn trees, brush, and even the surface of the moss, without altering the underlying permafrost.

    The occurrence of permafrost in peatlands in the southern part of the discon-tinuous zone appears to be related to changes in the thermal conductivity of the peat through the year (R. J. E. Brown 1966a, b, Tyrtikov 1959). During the summer the surface layers of peat become dry through evaporation. The thermal conductivity of the peat is low and warming of the underlying soil is impeded. The lower peat layers gradually thaw downward and become wet as the ice layers in the seasonally frozen layer melt. In the autumn there tends to be more moisture in the surface layers of the peat because of a decreased evaporation rate. When it freezes the thermal conductivity of the peat is increased considerably. Thus the peat offers less resistance to the cooling of the underlying soil in winter than to the warming of it in summer. As a result, the mean annual ground temperature under peat will be lower than under adjacent areas without peat. When conditions under the peat are such that the ground temperature remains below 32° F throughout the year, permafrost results and is maintained as long as the thermal conditions leading to this lower temperature persist.

Although the influence of the ground vegetation on permafrost is dominant, trees are of some importance. They shade the ground from solar radiation and intercept some of the snowfall in winter. The effect of even a single tree in shading the ground in summer and reducing the snow cover at its base in winter appears to influence the heat exchange at the ground surface sufficiently to produce a lens of permafrost beneath the tree (Viereck 1965). The density and height of trees influence the microclimatic effects of ground surface wind velocities. Wind speeds are lower in areas of dense growth than in areas where trees are sparse or absent. The movement of air represents the transfer of heat from one area to another. In peatlands the trees are generally stunted and scattered, and there are numerous open areas that permit higher wind speeds and the increased removal of heat per unit time. Therefore, the possibility of slightly lower air temperatures and ground temperatures, because of higher wind speeds, is greater than in areas of dense tree growth (Johnston et al. 1963). Investigations in temperate regions indicate that the nature of peatland surfaces contributes to air temperatures being several degrees lower than over other natural adjacent surfaces (Williams 1966). It is quite probable that the same situation prevails in permafrost regions.

Permafrost exerts considerable influence on the environment in which the subsurface organs of plants have developed (Tyrtikov 1959). The effects of permafrost are mostly detrimental to plant development because of its cold temperatures and impermeability to moisture. Permafrost impedes warming of the soil during the growing season and the temperature of the root zone is considerably below the optimum. Absorption of water by the roots is reduced which leads to physiological dryness of the plants. On the other hand, water is gradually released to the root zone through the summer as the active layer thaws. Root development is retarded and roots are forced to grow laterally because downward penetration is prevented by the permafrost. Many large trees cannot be supported by these shallow root systems and they lean against each other or fall over resulting in a "drunken" forest.

Permafrost forms an impermeable layer which impedes downward drainage leading to a decline in aeration and impoverishment of nutritive substances because of the weakening of the activity of micro-organisms. Soil movement in the active layer also influences the vegetation. Frost action causes unevenness in the ground surface, solifluction and other downslope mass movements of earth material disturb the vegetation, and thermokarst changes the surface configuration of the ground – all producing detrimental influences on the vegetation. These various influences of permafrost on vegetation increase northward as permafrost becomes increasingly widespread and the active layer becomes thinner.

c / Drainage

Water greatly influences the distribution and thermal regime of permafrost. In the discontinuous zone, the existence of permafrost is inhibited in poorly drained areas. Precipitation influences the depth of thaw and soil temperatures (Shvetsov and Zaporozhtseva 1963). First, the amount of moisture in the soil immediately before it freezes in the autumn determines the ice content and depth of thaw the following summer; second, the moisture content of the soil surface and the infiltration of atmospheric water influence the heat transfer to the frozen soil during the thaw period. Moving water is an effective erosive agent of perennially frozen soils. There is almost always an unfrozen layer of ground under water bodies that does not freeze to the bottom. The extent of this thawed zone varies with a large number of factors – area and depth of the water body, water temperature, thickness of winter ice and snow cover, general hydrology, and composition and history of accumulation of bottom sediments (R. J. E. Brown 1966a, b, W. G. Brown et al. 1964, Johnston and Brown 1964, Johnston and Brown 1966, Lachenbruch 1962). The ocean has an important thermal influence on permafrost causing it to be thinner at the shore than inland (Lachenbruch 1957).

Permafrost greatly influences the hydrological regime. Its impermeability to water is responsible for the existence of many small shallow lakes and ponds in the continuous zone and in the northern part of the discontinuous zone where permafrost is widespread. Beaded streams are another indication of the influence of permafrost. Irregular enlargements of stream channels result from the melting of masses of ground ice beneath streams as shown in Figure 13.

d / Snow cover

Snow cover influences the heat transfer between the air and the ground and hence affects the distribution of permafrost. The snowfall regime and the length of time that snow lies on the ground are critical factors. A heavy fall of snow in the autumn and early winter inhibits winter frost penetration. On the other hand, a thick snow cover that persists on the ground in the spring delays thawing of the underlying ground. The relation between these two situations determines the net effect of snow cover on the ground thermal regime. In the discontinuous zone, particularly in the southern fringe, it can be a critical factor in the formation and existence of permafrost. For example, the thickest permafrost in the southern fringe of the permafrost region occurs in palsa on which snow cover is thin because of their exposure to wind (Lindqvist and Mattsson 1965, Sjörs 1959a, b, 1961). In the continuous zone snow cover influences the thickness of the active layer.

The considerable influence of snow cover on the ground thermal regime can best be illustrated by several quantitative examples. At Norilsk, USSR (mean annual air temperature 16.9° F), it was shown that a snow cover exceeding five feet completely damped out air temperature influences on heat emission from the ground (Shamshura 1959). Studies at Schefferville, PQ (mean annual air temperature 23.9° F), indicated that snow cover is a dominant factor in controlling permafrost distribution at that site (Annersten 1964). Variations in snow cover caused temperature variations greater than those resulting from vegetation cover. It was postulated that a snow depth of about sixteen inches could be regarded as the critical snow depth for permafrost to survive. Beneath a greater depth either no permafrost existed or a degrading condition prevailed.

e / Fire
Fire is a transient factor which is not normally considered as affecting the permafrost. Although the time of actual burning at any given point is usually of short duration, a forest or tundra fire can have a marked influence on the ground thermal regime. Many fires started by lightning occur every year in the permafrost region, particularly the discontinuous zone which is mostly forested. Through the years much if not all of this part of the permafrost region has been burned over more than once. Tundra fires are not as common because of the wet ground conditions and low proportion of woody material in the vegetation complex. Fires occur frequently, however, in the spruce-lichen woodlands of the forest-tundra transition where the thick dry continuous lichen mat may burn or smoulder for many months during the summer.

The degree of influence of a fire on the permafrost depends on the condition of the vegetation and the rate of burning. A fire may move rapidly through an area burning trees but only charring the top surface of the ground vegetation. Palsas and peat plateaus have been observed in the Hudson Bay Lowland and Northwestern Manitoba where the trees have been burned and the surface of the moss and lichen charred by the fire. Below a depth of one inch this vegetation layer was untouched, its insulating effect on the underlying permafrost unaltered from nearby unaffected areas which did not catch fire (R. J. E. Brown 1965b and 1968b).

If dry conditions have prevailed in an area for some time prior to a fire, considerable change in the permafrost may occur. For example, a fire began on August 11, 1968 in the vicinity of Inuvik NWT located in the forested part of the continuous permafrost zone. Summer rainfall had been unusually light and even the normally moist moss cover was dry. The fire swept through flat lands and hillsides with such intensity that even frost mounds, which occur extensively throughout the area, were denuded of all vegetation. The removal of the insulating surface cover exposed the ice laden perennially frozen soils to thawing. Water from melting ground ice on the fire-bared hillsides northeast of Inuvik was released in sufficient quantities to cause considerable thermal erosion during the remainder of the thawing season to mid-September. Firebreaks were bulldozed to the permafrost table. In less than one month, water courses several feet wide had eroded eighteen inches into the ice laden permafrost. In the following summers deeper thawing and thermal erosion of the permafrost will continue because of the absence of the surface vegetation. The permafrost table will eventually stabilize as a new thermal equilibrium is reached and the vegetation gradually regenerates but the effects of the fire will be felt for many years (Watmore 1969).

f / Soil and rock
Bare soil and rock have considerable influence on the temperature of the ground because of their ability to reflect solar radiation. Reflectivity values in the range of 12 to 15 per cent for rock and 15 to 30 per cent for bare soil have been observed. There will also be different evaporation rates and intakes of precipitation. Variations in thermal properties such as conductivity, diffusivity, and specific heat affect the rate of permafrost accumulation. The thermal conductivity of silt, for example, is about one-half that of coarse-grained soils and several times less than that of rock (Kersten 1966). These factors assume their greatest significance in the continuous zone where the climate is sufficiently cool to produce permafrost regardless of the type of terrain (Brown and Johnston 1964).

The influence of permafrost on soil and rock is manifested by such phenomena as solifluction, and other downslope mass movements of earth material. These movements and frost action in the active layer tend to break down coarse soil particles and rock fragments into fine-grained material. Intensive frost action is also responsible for the heaving of massive blocks of fractured bedrock.

g / Glacier ice
The growth and regime of glaciers and ice caps is determined by climate but ice is considered as a terrain factor because like vegetation, water, and snow it forms a layer on the ground surface between the permafrost and the atmosphere which affects the heat exchange between them. It has been postulated by many workers

that the bottom temperature beneath much of a continental ice sheet is below 32° F. In temperate glacier conditions, the ice bottom temperature is at the pressure melting point. Beneath an ice sheet 3000 feet thick, for example, the temperature of the water film at the bottom of the ice would be about 31.5° F. In polar glacier conditions, the bottom of the ice is frozen to the underlying ground and the temperature at the ice-ground interface is below 32° F. Both of these glacier conditions probably occur extensively throughout an ice sheet. Consequently, beneath continental ice sheets in the northern hemisphere permafrost was probably widespread but thin because of the proximity of bottom temperatures to 32° F. Permafrost may have been somewhat thicker beneath the margins of ice masses where the effect of cold air temperatures can penetrate to the underlying ground (Shumskiy 1964). After the ice retreated, permafrost in areas covered by post-glacial inundations would dissipate and would not have re-formed until these bodies of water receded several thousand years later. In contrast to areas covered by ice sheets, much colder temperatures were imposed by the periglacial climate on ice-free areas – producing thicker and colder permafrost.

Thus, continental glaciation has undoubtedly exerted great influence on permafrost conditions from one region to another. It is notable that ice sheets were much more extensive in the western hemisphere than the eastern hemisphere. All of Canada, excluding the western Yukon Territory and possibly the northwest part of the Arctic Archipelago, was covered with ice sheets during the Pleistocene (Flint 1957). In contrast, a large part of central and eastern Siberia was not covered and here continuous permafrost now extends more than 500 miles south of the treeline, and the thickest permafrost in the northern hemisphere has been recorded.

h / Conclusion

Fluctuations have occurred through time in the extent, thickness, and temperature of the permafrost in response to changes in climate and terrain. Since its initial formation, the permafrost in any area may have dissipated and re-formed several times during periods of climatic warming and cooling. Glacial history has had a marked effect. Changes in vegetation caused by fire, climatic succession, encroachment in water basins, or by the permafrost itself all have pronounced local effects. The regime of the fall and accumulation of snow influences the ground thermal regime. The geothermal gradient also affects the ground thermal regime. It varies in different types of soil and rock, with changes in geological structure and with time.

Thus, the environment in which permafrost exists is a complex dynamic system, the product of past and present climate and terrain features, which are in turn influenced by the permafrost. The thermal sensitivity of permafrost is such that even small changes in climate and/or terrain will produce changes in the extent, thickness, and temperature of the permafrost. The interactions of permafrost and these factors in the North are varied and very complex. Even a slight change in one factor produces a change in one or several other factors.

D PHYSIOGRAPHIC REGIONS

Permafrost occurs in five physiographic regions, the Canadian Shield, Hudson Bay Lowland, Interior Plains, Western Cordillera and Arctic Archipelago (see Figure 3). Extensive descriptions of these regions are available in several publications from which the following brief accounts are given of their soils and permafrost characteristics and related engineering considerations (Bostock 1964, Harwood 1965, Lang 1961).

The Canadian Shield, covering about two million square miles, is the largest physiographic province. It extends in a great arc around Hudson Bay and a large portion of it lies within the permafrost region. The terrain consists of rock knobs interspersed with innumerable lakes and poorly drained depressions. Soil cover on the rock knobs is generally thin (or absent) and consists of glacial deposits, lake and marine silts and clays. The same soils occur in depressions and are commonly overlain by peat. In the southern fringe of the discontinuous zone permafrost islands occur in the better drained portions of bogs and peatlands. Northward, the permafrost increases in extent and in the continuous zone it is found in the bedrock. The intensive glaciation of this physiographic region has resulted in a lack of available earth materials in many areas as a source for building foundation pads and fill for roads and other construction. Much of the available soil is fine-grained and frost-susceptible and contains considerable quantities of ice.

The Hudson Bay Lowland is a low flat area and beach ridges marking the limits of postglacial marine submergence are the only major relief features. Drainage is poor between river valleys. Soils consist of thick peat deposits overlying marine sediments and till. Local microrelief features include spruce islands, palsas, and peat plateaus. The lowland lies within the discontinuous zone except for a narrow strip along the Hudson Bay coast in the continuous zone. In the discontinuous zone, permafrost occurs in scattered islands mostly in the microrelief features. Suitable earth materials for construction are scarce except in the beach ridges which are the source of large quantities of granular coarse-grained soils suitable for buildings and roads. The thick peat deposits extending over vast distances make road construction difficult and the peat plateaus and other microrelief features hinder trafficability of offroad vehicles.

The relief of the Interior Plains which includes the Mackenzie River valley is rolling with isolated highlands. Soils are predominantly fine-grained. In the southern fringe of the discontinuous zone, permafrost occurs in scattered patches in peatlands; further north it becomes more widespread. Only the extreme northern portion lies in the continuous zone. It appears that permafrost does not exist beneath the Mackenzie River or water bodies in its delta. As in the Canadian Shield, the fine-grained soils which predominate are frost-susceptible and contain large quantities of ice. Suitable sources of earth materials for construction are scarce in many areas.

The mountainous Western Cordillera area consists of ranges, plateaus, and intermontane valleys and trenches. The distribution of permafrost is complicated greatly by the relief as described. The soils are extremely variable, earth materials suitable for building purposes being plentiful in some areas and lacking in others.

The Arctic Archipelago, comprising the Arctic Lowlands, Innuitian Region, and the northern part of the Canadian Shield, lies entirely within the continuous permafrost zone, except possibly for the southeast tip of Baffin Island. The active layer is thin and the permafrost is hundreds of feet thick. All earth materials are frozen and have to be thawed before they can be used for construction. Some areas of the Archipelago have extensive glaciofluvial deposits which are good sources of building materials. Large masses of ground ice have been encountered even in these coarse granular materials which can be a hazard to the satisfactory performance of engineering structures.

2 Engineering Considerations

Although frozen soil provides excellent bearing for a structure, its strength properties are greatly reduced with increase in temperature and, if thawed, may be lost to such an extent that it will not support even light loads. The most serious difficulties arise with those soils, usually fine-grained, that have large moisture (ice) contents. When thawed these materials turn to a slurry with little or no strength, and large settlements and perhaps failure of a structure may occur. Another factor to be considered is frost action in the active layer, which freezes and thaws seasonally. The active layer often consists of frost-susceptible soils and, in addition, is saturated with moisture. A ready supply of water, which is a prerequisite for frost heaving, is thus provided. Differential heaving as a result of frost action can cause serious damage to foundations and buildings. Satisfactory performance can be achieved by reducing the depth of frost penetration or replacing such soil with material in which ice accumulation will be minimal.

Three features of permafrost are significant in engineering construction: (1) Permafrost is particularly sensitive to thermal changes. Any natural or man-made change in the environmental conditions, however slight, will greatly affect the delicate natural thermal equilibrium (see Figure 14). (2) Permafrost is relatively impermeable to moisture. Drainage is vital, therefore, because all movement of water occurs above the permafrost; in northern areas surface water is conspicuous, despite the generally low precipitation. If natural drainage is impeded, or proper drainage structures are not provided, construction operations can be seriously complicated by intensified frost action during the winter and accelerated thawing during the summer. (3) The ice content of frozen ground is a most important consideration. Solid rock, gravel, and sand usually contain little or no ice at temperatures below 32° F and few difficulties are encountered in building on these materials. Most problems arise with fine-grained materials and organic materials (such as peat) which usually have extremely high ice contents and are susceptible to frost action. As long as the water remains frozen in such soils, the ice binds the individual particles together to produce a material with considerable strength; when thawed, however, these soils can change to a soft slurry with little or no strength.

Experience with permafrost and failures attributed to its unusual properties have shown that it affects every type of engineering project, be it selection of a new town-site or erection of a single structure in an established community. Even though much information has been collected in recent years on the occurrence of permafrost and its properties, the need for adequate site investigations cannot be overemphasized. The form and extent of investigations may vary, depending on particular job requirements, but certain basic procedures should be followed in acquiring the information needed for good design and construction. The following account is derived mainly from the excellent paper on engineering site investiga- tions in permafrost areas by G. H. Johnston (1966b) which reviews all aspects of this important topic.

Site investigations in any region are usually conducted in three phases, namely: preliminary office studies and planning, field investigations, and laboratory and office studies (preparation of reports). Planning of each phase requires greater attention when northern sites are considered, primarily because of permafrost but also because of the relative isolation and general lack of knowledge of the terrain.

"Site investigation," as used here, is limited to those phases of an over-all site evaluation which pertain to the collection of information on permafrost condi- tions. The procedures described apply to all types of engineering projects rather than only to specific types of projects, such as buildings, water supply and distribu- tion systems, or road location and construction.

At least a year is generally required to complete an adequate engineering site investigation in permafrost areas. Some overlapping of the three phases will occur since certain portions of each are better conducted at different times of the year. Depending on the type of project, investigations may be completed in much less than one year; for others, work may continue well into a second year, but in general, a one-year period is needed to assess properly all site conditions affecting design and final planning (Muller 1945).

1 *Preliminary Office Studies and Planning*

Emphasis must be placed on careful planning and compilation of pertinent infor- mation prior to the conduction of field investigations. This preliminary work is normally carried out several months prior to the field season. For most areas, the season begins in March or April. Much time and effort can be saved, and survey costs reduced, if the area has been thoroughly studied beforehand. In addition, depending on information acquired during this initial study, some assessment can be made of foundation designs and construction techniques which might be used.

a / Background information

Although meagre information is available for many northern regions, data on adjacent or similar areas may provide useful information on general conditions which may be encountered. Of particular interest is a knowledge of the geology and climate; both are intimately connected with the formation and existence of permafrost.

Many northern areas have been geologically mapped, albeit superficially in most instances. Reference to such maps and to published reports can provide much information of value. Most of the Arctic and Subarctic have been glaciated; thus, accounts of the glacial history are of specific interest with respect to soil and permafrost conditions. Accounts of the travels of many early explorers and geologists provide valuable descriptions of the terrain and geomorphology (Leffingwell 1919).

Climate plays an important role with respect to permafrost particularly in the ground thermal regime (Legget *et al.* 1961). Broad correlations exist between the occurrence of permafrost and air temperature (R. J. E. Brown 1966c, Thompson 1966). Meteorological records should always be examined and summarized to evaluate local climate in relation to permafrost. Knowledge of local climate is also invaluable, of course, in planning field operations.

A complete folio of pertinent maps, geologic, topographic, and hydrologic (if available) covering the areas to be investigated, will provide further information and will also be useful for recording field observations. Scales of 1:50,000 or larger are most useful but maps of smaller scale should not be disregarded. Most of northern Canada has been mapped to some degree and maps are generally available through the federal government.

An important source of information, often neglected, is contact with agencies, firms, or individuals who work or have worked in the areas under investigation. In many cases, useful site information is available from those with exploratory, development, or construction experience in the area, such as mining companies, trappers, and prospectors. Contact with such people can be made prior to and during the field exploration. Observations on performance of existing structures are also most valuable.

b / Airphoto studies

Airphotos provide the most valuable aid for preliminary planning and site evaluation (Frost 1960). Airphoto coverage, to various scales, is now readily available for most northern regions. The airphoto serves as a map; the surface features on this map (together with a knowledge of the geologic and climatic history of the area), can when properly interpreted, yield a wealth of information on subsurface conditions (Frost 1952, Belcher 1948). Soil types and permafrost conditions are indicated by or can be inferred from relief, vegetation, and drainage characteristics. Photographic tones provide further clues. Detrimental permafrost areas can be identified and delineated by recognizing phenomena such as soil polygons and patterned ground forms resulting from frost action (Washburn 1956). Ice wedge polygons may vary in diameter from 15 to 500 feet (Figure 7). Pingos (Figure 8), thermokarst lakes, drunken forests, or ground subsidence caused by thawing of large buried ice masses, solifluction lobes or terraces, frost mounds, hummocks, and mud boils can all be readily identified; when seen on an airphoto, they indicate potentially unsuitable foundation conditions.

The topographic position of a locality is perhaps the most important feature related to permafrost that can be recognized on an airphoto; this assists greatly in predicting permafrost conditions since detrimental permafrost conditions generally occur in certain topographic situations. For example, many of the phenomena just mentioned are associated with low-lying areas, such as coastal plains and stream valleys or depressions in upland areas.

Dense forest growth, particularly in the discontinuous permafrost zone, may mask ground surface features on an airphoto. Certain vegetation associations are closely related to subsurface conditions and permafrost occurrence and therefore, when considered together with topographic position, soil texture, and drainage, can serve as fairly reliable indicators (Britton 1957, R. J. E. Brown 1966a, Hansen 1953, Hustich 1953).

In general, the surface drainage pattern is little altered by permafrost. However, local drainage patterns and characteristics, such as beaded streams (as shown in Figure 13), seen on airphotos do provide useful clues. Water, either still or running, has a thawing effect on permafrost because of its heat storage capacity and powers of erosion. Slumped lake banks and polygonal surface features disintegrated by running water are evidence of this effect.

Most surface features associated with permafrost are the result of a complex relationship of many factors such as climate, geology, freezing and thawing, relief, and drainage. Much depends on the experience of the airphoto "reader" and his ability to analyze and interpret what he sees. Even though permafrost and general terrain conditions can be predicted fairly reliably from airphotos, it is still necessary to visit selected areas which have been subdivided on the basis of similar terrain characteristics, such as relief, drainage, and vegetation, in order to verify the interpreted conditions (Johnston *et al.* 1963, Sager 1956).

c / Planning
Careful planning for field operations is essential if the work is to proceed efficiently and economically. Movement of personnel and equipment into and about the areas under investigation is affected greatly by difficult access and lack of transportation

at certain times of the year. Special care is required in planning for projects, such as road location surveys, which cover large areas and varying types of terrain. Every detail must be carefully thought out because the field season is short and supply lines are usually long and difficult. Good communications are essential.

Personnel selection is very important. An experienced civil engineer should supervise and co-ordinate all activities. Most surveys require a glacial geologist preferably with training in geomorphology, a soils engineer, and/or civil engineer assisted by technicians. All should be well qualified and very familiar with permafrost. Depending on the objectives and scope of the project, other specialists may be required, such as a botanist and a hydrologist. If local labour is not available, additional assistance will have to be brought to the site.

The success of the investigation will depend largely on the degree to which the work has been preplanned and on the knowledge of the area which the field workers have acquired before entering the second phase, that is, the field investigation.

2 Field Investigations

Field investigations are usually carried out in two stages: an exploratory survey is conducted over wide areas to assess general site conditions for detailed examination; and thorough investigations are conducted at selected sites to gather detailed information on permafrost conditions relevant to the design of structures and the construction techniques to be used.

a / Exploratory survey

Potential sites or routes selected by preliminary office studies, primarily through the use of airphotos, are examined during this stage. These areas are evaluated for their suitability in all seasons. Geological studies include bedrock control and the glacial history, with which permafrost is closely associated, of the over-all area with more detailed examinations at specific locations.

A terrain reconnaissance is made to map topographic features including relief, drainage, and vegetation patterns and characteristics. Areas of patterned ground and permafrost phenomena are delineated. Locations of icings are noted. Lake and stream ice thicknesses and snow accumulation patterns and depth are observed. Transit and tape and level surveys are made to establish horizontal and vertical control for the area. Field sketches or plans are prepared on which all terrain information can be recorded. Data can also be noted on airphotos or topographical maps already available.

Determination of the distribution of permafrost is the most important aspect of the exploratory survey. Selected areas are examined by borings, test pits, and probings to check predictions and to determine actual subsurface conditions.

Distribution of permafrost, its areal and vertical extent, and factors which appear to control its existence such as drainage (surface and subsurface), vegetation (type and thickness of moss cover), and topographic position are very important, particularly in the southern fringe of the permafrost region. The depth to which seasonal freezing and thawing penetrates (active layer) and the rate at which these processes take place, the depth to the permafrost table, the presence of taliks within permafrost, and the movement of subsurface water are also factors that must be determined. Types of ice segregation and various materials with which they are associated must be known and delineated.

Samples of perennially frozen materials can be obtained in several ways (G. H. Johnston 1963b). Naturally occurring exposures and hand boring methods will provide general information for relatively shallow depths. Test pits and core drilling provide detailed information on soils and ice segregation to greater depths.

Geophysical methods to determine the depth and extent of permafrost have had increasing application (Barnes 1966, Hatherton 1960, Hobson 1962, Joesting 1954, Roethlisberger 1961). Seismic refraction soundings appear to be suitable for determining the upper limit of, or the depth to, permafrost; electrical resistivity methods are most useful in determining the thickness of permafrost. Experience in their use is limited, however, and reliability of results depends to a great extent upon a careful interpretation of data.

Suitable instrumentation and observation programs to obtain data for the designer and planner should be set up early in the exploratory stage of the field work. Meteorological observations, including measurement of air temperatures, precipitation (rain and snow), and wind (direction and velocity) should be made to provide information on local climate. Permafrost is defined on a temperature basis; thus, data on the ground thermal regime is needed to describe permafrost conditions. Ground temperature installations should be made to depths of at least twenty feet and greater if possible. Ground movement gauges may be necessary in many cases to determine detrimental effects due to freezing and thawing of the active layer. Observations on the rate and depth of thaw under various terrain conditions are also essential and should be made throughout the thawing season (the maximum thaw will be observed in the late fall) by simple hand probing methods. To provide useful information, such observations should be made on a regular daily, weekly, or monthly basis; since, in most cases, the observation period should cover at least six months, the observations should be started as soon as possible.

Finally, the construction and performance history of similar structures in the area will provide much information on site conditions. These sources will prove of inestimable value to the designer and should not be overlooked. Observations on existing structures should be initiated at this stage.

Exploratory survey is the most important stage of a field investigation. Although its purpose is to obtain general information on surface and subsurface conditions over a wide area, and it might therefore be considered rather superficial, results obtained dictate to a large extent the future of the proposed project. On the basis of the information obtained, unsuitable areas are eliminated and final sites and routes selected for more detailed examination. Design criteria and construction methods and techniques also evolve during this stage. Thus, it is a critical period of extensive and intensive examination of all factors related to permafrost and its effect on the proposed project.

b / Detailed investigation
This phase of the field program provides the detailed information needed for final planning and design of the project. Various construction methods and techniques are assessed and selected during this stage. This work is done at sites selected during the exploratory survey to supplement information already obtained. Preliminary designs may be drawn up during this stage and layout of structures at approved sites begun. The field supervisor of this stage should be completely familiar with the requirements of the project so that he can direct the field program along the right paths. Close co-ordination of all operations between field and head offices is necessary.

Observations begun during the exploratory survey, including depth of thaw in disturbed and undisturbed areas and ground temperature measurements, are continued and expanded. More detailed records of subsurface conditions at actual construction sites are required in selecting foundation designs (Nees and Johnson 1951). In particular the ground thermal regime should be critically analysed, and the form and extent of ice segregation in the underlying materials should be noted in detail (G. H. Johnston 1963b).

Test pits have particular application to site exploration in areas covered with deposits of stony tills or gravel. A major advantage of this method is that it permits the frozen soil and the ice segregation to be examined in the undisturbed condition. They can be excavated at any time of year to depths of 20 to 30 feet by compressed air or gasoline engine jackhammers. Core drilling methods to obtain undisturbed frozen samples are widely used for investigations to depths of 20 feet or greater (as shown in Figure 16). Although coarse-grained soils have been successfully sampled using special refrigeration equipment and techniques (Hvorslev and Goode 1957) drilling methods are most applicable in fine-grained soils. Good cores of undisturbed material can be sampled for moisture (ice) content and unit weight determinations and for identification and classification of soils encountered. Some testing may be done at the site, but many samples are shipped out in plastic containers (to reduce weight and thus transportation costs) for laboratory analysis. Although refrigerated methods can be used to ship samples, it is difficult to preserve specimens in the frozen state and therefore photographic techniques have been developed to provide a permanent record.

At this stage actual test installations at the site can be made and construction procedures can be developed. For instance, if pile foundations are under consideration, field studies might include an evaluation of pile-placing techniques and pile load and pull out tests to determine adfreezing strengths. Test fills might be constructed for road and airstrip design purposes. Bearing capacity tests of frozen soil might also be included. Although it takes time to accumulate useful results, much valuable information can be obtained at this stage, particularly for large construction programs which may take several years to complete, such as roads, railroads, and townsite developments. Some construction may be started during this early period but it will be generally limited to such activities as site preparation and opening of borrow pits. In these cases it is useful to observe methods of excavating and the handling and placing of frozen and thawed materials.

Detailed terrain conditions are accurately and specifically described and delineated. Topographic maps, with contour intervals of from two to five feet are

FIGURE 16 Drilling in permafrost near the Mackenzie River delta in continuous permafrost zone to determine soils and permafrost conditions. The drill can be dismantled and transported to another location by helicopter.

required to portray surface configurations and conditions. Subsurface information including soils and permafrost conditions at or along finally selected sites and routes are shown on plans, and sections or logs and are described by test results and written reports.

During both the exploratory and detailed surveys, information collected in the field is sent back to the head office for evaluation with respect to the project as a whole. Although over-all direction of the field work may come from there, many decisions as to the course of the work must be made by the field supervisor. Various phases of each type of survey may overlap or be carried on at the same time and in conjunction with each other. Nothing should be overlooked or omitted during these stages so that final planning and design may proceed without delay. It may not be possible to obtain missing or forgotten information until the following year.

3 Final Studies and Preparation of Reports

This phase of site investigation consists primarily of office studies directed toward presentation of all site information needed for the planning and design of engineering structures. Evaluation of laboratory test results of the physico-mechanical properties of the materials encountered at construction sites are of prime importance for foundation design and selection of construction procedures and techniques. Preparation of detailed maps and drawings showing terrain conditions at appropriate scales is necessary for final route selection and proper siting of structures. The designer and planner is dependent wholly on information in the final reports and recommendations of field workers. Emphasis must be placed upon preparation of thorough and detailed reports.

Some consideration should be given to continuing and expanding observations begun during field investigations. Observations, such as ground temperatures, depth of thaw in undisturbed and disturbed locations, and effects of and changes in drainage patterns (both surface and subsurface) though not always of immediate value will provide useful information for future projects in the area. Studies should be continued by instrumentation and observation of the performance of various structures and their effects on permafrost conditions. Such studies will provide valuable information for future work of similar nature.

Many investigations have been made at various northern locations. Unfortunately, experience gained and conditions encountered are seldom recorded. There is a great and immediate need for such information to increase the knowledge of permafrost and the conditions under which it exists. All engaged in northern work are urged to record their observations for the benefit of improving construction techniques and performance.

B   DESIGN AND CONSTRUCTION

When general site conditions have been evaluated, further detailed investigations are normally required at the locations of individual structures. The results of these will indicate the approach to be taken in foundation design and the construction techniques to be used. These are usually considered under one of the following headings:

1
Neglect of permafrost conditions.
2
Preservation of frozen conditions for the life of the structure.
3
Elimination of frozen condition or material before building.
4
Thawing of frozen ground which will occur during life span of structure with expectation of subsequent ground settlement; foundation design which takes the expected movement into account.

Permafrost can be neglected when engineering works are sited on well-drained coarse-grained soils or bedrock; conventional design and construction procedures can then be used. Where fine-grained soils with high ice content are encountered in the zone of continuous permafrost, every effort must be made to preserve the frozen condition. In the discontinuous zone, it may be convenient to remove such materials by thawing or excavation and to replace it with well-drained material not susceptible to frost action; standard foundation designs can then be used. For some structures, in either the continuous or discontinuous zone, it may not be possible to prevent thawing of the ground during the life of the structure and settlement must therefore be anticipated and taken into account in the design.

Preservation of the frozen condition can be accomplished by either ventilation or insulation construction techniques; the former is commonly used with heated buildings. Foundations are well embedded in the permafrost, and the structure is raised above the ground surface to permit circulation of air beneath to minimize or prevent heat flow to the frozen ground. Pile foundations placed in steamed or drilled holes have proved well suited to this method, and have been used extensively in northern Canada and elsewhere (Pihlainen 1959).

Where pile placing may be difficult, as in very stoney soils, alternative foundation designs may prove more economical. Insulation to prevent or reduce thawing of the underlying frozen material may be achieved by placing a gravel blanket on the surface of the ground on which the structure is to be erected. This method is generally limited to small low-cost buildings that can tolerate some movement.

For the construction of highways, railways, and airstrips (Linell 1957), where the ventilation technique cannot be applied, the insulation method must be relied on. Normally, fill methods are used throughout, and disturbance of the ground surface cover is kept to a minimum. Cuts through hills are avoided where possible. Proper drainage must be provided to prevent accumulation of water, which would thaw the underlying permafrost, and the formation of icings, which can block a road during the winter. The procurement of large quantities of fill for road and airstrip construction presents its own problems; suitable sources must be located, cleared of vegetation, and allowed to thaw well in advance of construction operations.

Excavation of frozen ground can be difficult and costly because normal excavation techniques are much less effective in permafrost. For some structures, however, it may nevertheless prove economical to excavate the frozen soil, replacing it with material not susceptible to frost action on which the foundation can be built. This method is particularly applicable in the southern fringe area. Again, adequate drainage must be provided to take care of seepage water.

Permafrost complicates the provision of water and sewer services (Dickens 1959). Only limited year-round sources of water are available because many lakes and streams freeze to the bottom during the winter, and water-bearing strata are only occasionally encountered in permafrost. Normal methods of sewage disposal into the ground are generally prohibited because of the imperviousness of the permafrost. Distribution systems are generally located in insulated boxes (utilidors) on or above the ground surface because of problems resulting from excessive thawing of the frozen ground caused by heat loss from the pipes or, conversely, freezing of pipes if they are placed in the active layer or in the permafrost.

The thawing effect of water on perennially frozen ground becomes particularly critical when dams and dykes are constructed on permafrost and large areas are covered by impounded water (Johnston 1965). The rate at which thawing will take place and the depth to which thaw will penetrate beneath the water and the water-retaining structures are of prime importance in the design of their foundations. In some cases the underlying frozen ground can be excavated and the structure placed on bedrock. The frozen condition can be retained by natural or artificial refrigeration; or the embankment can be built up as settlement occurs when the permafrost thaws.

Permafrost has varying effects on mining operations depending on the mineral and mode of exploitation. In subsurface hardrock mining, water seeping into shafts and drifts may freeze and fill these openings with ice, thus hampering normal operations. Perennially frozen placer deposits and unconsolidated ore deposits containing ice require thawing and special excavation techniques which increase costs. In oil drilling difficulties may arise where rigs have to be located on fine-grained soils containing ice. Drilling for and bringing the crude oil to the ground surface through thick perennially frozen strata may be hampered by the sub-freezing ground temperatures. The construction and operation of pipelines can cause thawing of the permafrost and unstable ground conditions.

Because of permafrost the design, construction and maintenance methods, or techniques of many engineering structures have to be altered or modified to ensure adequate performance: buildings, municipal services, roads, runways, railroads, towers, and dams all require special consideration; economic activities such as mining and agriculture are influenced in a variety of ways toward modification of standard practices and introduction of new methods to cope successfully with this phenomenon. Permafrost is only one of several factors hindering human activities in northern Canada. However, an understanding of it is vital to the growth of what is at present a region of limited development.

3 Development of Permafrost Investigations and Northern Settlement

A HISTORY OF PERMAFROST INVESTIGATIONS

1 *Developments to Second World War*

Early references to unusual occurrences of frozen ground in the Canadian North are found in the records of early explorers. The first known mention of this phenomenon was made by Martin Frobisher in the latter years of the sixteenth century. Between 1576 and 1578 he organized and led three expeditions to southern Baffin Island to find a northwest passage. Attempts were made to found a settlement, and mine ore considered to contain gold. It was during the excavation of ore that frozen ground was encountered and recorded in his accounts.

In his review of the history of permafrost investigations in North America, R. F. Legget presented the following early references (Legget 1966). Joseph Robson, one of the early writers on the North, observed, in his *Account of Six Years Residence in Hudson's Bay from 1733–6 and 1744–7* published in London (Robson 1752), that: "The garden-ground at York fort and Churchill river thawed much sooner and deeper in the space of one month than the waste that lies contiguous to it ... by the heat therefore which the earth here would acquire from a general and careful cultivation, the frost might be soon overcome, that the people might expect regular returns of seed-time and harvest."

This suggestion is still valid for the southern fringe of the permafrost region but could not have been successfully applied in the Churchill area. He relates that "in September 1745 [he] tried the frost in the ground, by digging in a plain near the fort." The explanations he applied to what he observed make strange reading today, being based on the concept that "it is the moisture that communicates the freezing quality."

Another and more accurate observer of the eighteenth century in the North was James Isham. His "Observations on Hudson's Bay 1743" contained comments on frozen ground:

The shortness of the summer's is not Sufficient to thaw the Ice the severity of the winter occasion's therefore itt geathers more and more Every year, for which Reason

the frost is never out of the ground, in these parts, for in Dig'ing three or four foot
downe in the ground in the mids't of the summer, you shall find hard froze'n Ice, which
Ice may be ab't two feet thick, then come to soft ground again, for a small Depth and
above six or Eight foot Downe itt's all hard ice – in Summer it's with much Difficulty
you may Dig so Low Down.

Since the buildings used by the early settlers in the North were generally of
relatively simple construction, the perennially frozen nature of the ground upon
which they were erected occasioned little comment. Wooden sills placed directly
on the ground surface seem to have been the most commonly used type of foundation. Roads were non-existent apart from short stretches in settlements. Occasionally the frozen ground was disturbed such as for the burial of someone who
died in the North. Joseph Isham may be quoted again: " ... these men that are so
froze, or any one that Dyes and are Burried 6 foot under the surface of the ground,
continues hard froze for many Year's, and believe never will be thawed unless
taken up, – by the experience of frozen ground. ... "

No further observations on perennially frozen ground in northern Canada
were recorded during the latter half of the eighteenth century. Through the nineteenth century scattered observations were made in various parts of the North by
such explorers as Sir John Franklin and Sir John Richardson. In 1825 and 1826
at Fort Franklin on Great Bear Lake, Sir John Franklin reported a summer ground
thaw of only 22 inches (Richardson 1839). Sir John Richardson observed in
October 1835 that the soil at York Factory, at the mouth of Nelson River in northern Manitoba was thawed to a depth of 2 feet 4 inches, below which it was frozen
to a depth of 19 feet 10 inches. Two years later he excavated a pit 17 feet deep,
at Fort Simpson in October, revealing 11 feet of thawed soil above 6 feet of
frozen soil (Lefroy 1889). Traders of the Hudson's Bay Company, which had
already been in the Canadian North since the seventeenth century, and other
traders also made random observations. These observations were mostly measurements of the depth of thaw, incidental to investigations of other matters.

A notable exception to the general pattern was provided by the Royal Geographical Society, which was pre-eminent as a sponsor of scientific expeditions. In
the latter part of the last century, the Society had an active Committee on the
"Depth of Permanently Frozen Soil in the Polar Regions, its geographical limits,
and relations to the present poles of greatest cold." General Sir G. H. Lefroy, an
explorer, was Chairman, and in his report for 1889 presented observations on
permafrost for twenty-two locations, most of them in Canada. He included a
vigorous plea for greater efforts in investigating this natural phenomenon (Lefroy
1889).

The latter part of the nineteenth century also saw the first use of permafrost for
natural cold storage by American whalers at various points along the arctic coast.

There was a flurry of interest at the end of the century caused by the Klondike gold rush beginning in 1896 in Yukon Territory. The placer gold deposits were overlain by substantial thicknesses of perennially frozen gravel and organic material. Various techniques had to be devised to thaw the permafrost and remove the materials in order to extract the gold. A large number of scientific and popular papers and books were written on this subject which helped to attract worldwide attention.

In the twentieth century prior to the Second World War interest in permafrost was slight and references to it were few. Development of the petroleum reserves at Norman Wells, NWT, on the Mackenzie River, started in the twenties and here the detrimental results of thawing perennially frozen water-bearing silts and clays soon made themselves evident. Some experimentation was started here with the installation of some foundations on gravel pads. Valuable experience in construction in permafrost regions was gained also during building of the Hudson Bay Railway between The Pas and Churchill, Manitoba. This line was completed in 1929 and showed that construction techniques used in temperate regions can be modified successfully for permafrost conditions.

In the thirties gold mining began on Great Slave Lake and on Lake Athabasca, and the famous Eldorado uranium mine on Great Bear Lake was in production. All these mining ventures were in permafrost, but it was permafrost in solid rock almost exclusively and thus few unusual problems were encountered.

In the late thirties the Hudson's Bay Company moved away from its traditional practice of constructing small buildings with surface foundations in order to make a start at providing its post managers with modern improved living quarters. A new type of house was designed incorporating a concrete basement and standard furnace heating system. The first such house was built at Yellowknife on a site underlain by perennially frozen silty clay containing ice. Damage caused by thaw settlement caused by heat loss was so severe that the furnace had to be removed and the basement filled in.

2 Second World War and Postwar Developments

The Second World War caused rapid developments in northern Canada and stimulated interest in permafrost. The construction of the Alaska Highway to link the continental United States with Alaska by road and the associated construction of a network of permanent airfields comprising the Northwest Staging Route to provide an air link constituted the first major developments in northern overland transportation facilities. A similar network of airfields comprising the Crimson Route was built about the same time, through northeastern mainland Canada and Baffin Island. These were followed by the Canol Project involving the increased production of the Norman Wells oilfield and the construction of a pipeline and road across the Mackenzie Mountains to the Alaska Highway. The severe permafrost problems which were encountered publicized the need to study this phenomenon, for practical as well as purely academic reasons. Still farther to the north, the first Joint Arctic Weather Stations were established in the Queen Elizabeth Islands of the Arctic Archipelago in the immediate post-war period.

The imperatives of war allowed little time for experimentation or study and so most of the emergency northern construction of those years had to be carried out on a rather hit-and-miss basis. Considerable efforts were made to assemble useful information on northern building problems, the most notable contribution undoubtedly being S. W. Muller's book on permafrost, published in 1945 as *Permanently Frozen Ground and Related Engineering Problems*. This was the first complete English text on permafrost, being a compilation of information on fundamental and engineering aspects of permafrost obtained mostly from Russian sources.

Postwar activities in the north of Canada have been expanding continuously and interest in permafrost has increased apace. Because of the vast area of permafrost in Canada it was recognized that this phenomenon was a vital factor in northern development. The Division of Building Research, National Research Council of Canada, established a Permafrost Section in 1950 to carry out investigations on fundamental aspects of permafrost and engineering problems associated with it.

The first research project began the same year when the Division joined forces with the Directorate of Engineering Development of the Department of National Defence in sponsoring an expedition to the Mackenzie River valley in the Northwest Territories. The object of this small expedition was to obtain information on the construction and performance of buildings on permafrost from Great Slave Lake to the arctic coast. More than two hundred buildings were inspected, and their condition suggested the need for a critical review of site selection methods and the need for research on suitable building foundations in permafrost areas containing fine-grained soils with high ice contents, which are subject to extensive settlement on thawing (Pihlainen 1951).

Believing that the use of aerial photographs could simplify and reduce the costs of preliminary site surveys, the Division of Building Research joined the Defence Research Board and Purdue University in an expedition to the Northwest Territories during the summer of 1951. Purdue University, working under contract with the us Corps of Engineers, had five years' airphoto interpretation experience in Alaska and was anxious to extend its investigations to northern Canada. The findings of this expedition justified the belief in the applicability of airphoto interpretation methods for preliminary site surveys in permafrost areas.

These and other research needs, which could be fulfilled only by actual field investigations, prompted the Division to establish in 1952 a Northern Research Station at Norman Wells on the Mackenzie River about ninety miles south of the Arctic Circle. The choice of this site was prompted by the widespread occurrence of permafrost in the area, the transportation service by boat and air, and the co-operation offered by Imperial Oil Limited, which operates an oil refinery there. The operation of the research station was seasonal, extending from June to early October. After a decade of operation, this station was closed and replaced by use of facilities available at the Inuvik Research Laboratory operated by the Department of Indian Affairs and Northern Development.

Since 1950 the Division's permafrost research has been conducted along two broad fields of study (Brown and Johnston 1964): the first is concerned with acquiring knowledge of the nature of permafrost and its distribution in Canada, which are vital prerequisites to successful engineering; the second with the development of site selection methods and equipment, and building design and construction procedures and techniques in permafrost areas.

Research projects, many on a continuing basis, have been carried out at various locations in northern Canada. Information on the distribution of permafrost is being gathered continuously from a variety of sources including the technical literature, reports from others operating in permafrost areas, and direct field observations. A questionnaire was circulated in 1960 to government and private agencies in the North and considerable information was obtained on the distribution and occurrence of permafrost at many settlements.

Special attention has been given to the delineation of the southern limit of permafrost. Each year beginning in 1962 a survey of permafrost distribution and the factors affecting it in the southern fringe was carried out in a different part of Canada by road or helicopter with the aim of extending knowledge of the southern limit across the country (R. J. E. Brown 1964, 1965b, 1967a, 1968b). The last of six surveys, carried out in Labrador, completed this reconnaissance from the Atlantic Ocean to the Pacific (R. J. E. Brown 1969, in press). Consideration is now being given to the distribution of permafrost in the vicinity of the southern limit of the continuous zone. Accompanying this is a survey of the distribution of permafrost in mountainous regions of the western Cordillera. Several years of field observations including drilling and ground temperature measurements will be required to determine the pattern of permafrost distribution and occurrence.

Observations from anywhere in the permafrost region are recorded as they are received. With the several hundred observations of the occurrence of permafrost in northern Canada now available and plotted on maps, the location of the southern limit and the approximate division between the continuous and discontinuous permafrost zones are becoming increasingly evident (R. J. E. Brown

1960). A new separate permafrost map of Canada (in colour) was published jointly in 1967 by the Division of Building Research and the Geological Survey of Canada, Department of Energy, Mines and Resources (R. J. E. Brown 1967b).

Accompanying the collecting of permafrost information by these means is the continuing study of the basic factors affecting its distribution and continued existence in order to improve the ability to predict its occurrence. Such work involves investigations of the basic climatic and terrain components of the energy exchange at the earth's surface that affect the occurrence of permafrost (Legget et al. 1961). Field studies of some of these components were conducted at the Northern Research Station in 1959 and 1960, including measurements of evapotranspiration and radiation through the vegetative cover, and ground temperature beneath the cover (R. J. E. Brown 1965a). Closely allied with these studies is the continuing collection of ground temperature observations from thermocouple cables installed to various depths at scattered points across the permafrost region.

Since 1954 a major project has been the study of permafrost and engineering facilities at Inuvik, NWT, located on the east flank of the Mackenzie River delta, thirty-five miles east of Aklavik, NWT, in the zone of continuous permafrost. The building of this entirely new community was prompted in part by investigations conducted at Aklavik in 1953, which showed the soils and permafrost conditions in that area to be unusually poor from the construction standpoint (Pihlainen and Johnston 1954). The development of Inuvik (Pihlainen 1962) has provided a unique opportunity to observe the effect on permafrost of the construction of the various facilities associated with a town and the detailed assessment of the performance of the facilities themselves, including roads, airstrips, building foundations, and services. The initial phase of this project was concerned with the observation of construction procedures and the establishment of suitable instrumentation and reference points on the structures so that their performance could be observed and assessed in the future. The reference points were placed on all major structures to determine, by means of level surveys, the magnitude of any vertical movement that may occur; thermocouple cables were installed at various locations to measure ground temperatures.

The construction of the Kelsey hydroelectric plant on the Nelson River in northern Manitoba for Manitoba Hydro provided an opportunity to study dyke construction in an area of discontinuous permafrost. Beginning in 1958 thermocouple cables were designed, fabricated, and installed to observe the effect of the raised river level on the thermal regime of the permafrost, and settlement gauges were installed in the sand-fill dykes to record their performance. As expected, the water in the reservoir is changing the thermal regime of the underlying ground causing gradual thawing of the permafrost. The adverse effects of the accompanying settlement of the dykes are being accommodated by design and maintenance practices specially developed for this structure (G. H. Johnston 1965, 1969).

In 1960 a study of permafrost distribution and foundation problems was initiated at Thompson, Manitoba, the site of a new mining and smelting development of the International Nickel Company of Canada which uses the power generated at Kelsey. At the town, located in the southern fringe of the permafrost region, perennially frozen ground occurs in scattered patches and its temperature is close to 32° F. Construction operations are complicated because of the difficulty of predicting the occurrence of individual permafrost islands and also because disturbance or removal of the protective insulating moss cover generally causes the thawing of the underlying frozen soil containing ice layers, resulting in large-scale settlements of structures. Because of the Division's interest in determining the climatic and terrain factors affecting the distribution of permafrost and in studying construction problems encountered in the southern fringe of the permafrost region, of which the Thompson area is typical, an observation programme is continuing (Johnston *et al.* 1963).

The studies at Kelsey led to the realization of the paucity of information concerning the magnitude of the thawing effect of water in contact with permafrost. As a first approach and in an attempt to provide some information on this effect, a drilling programme was carried out in April, 1961, to determine the present level of permafrost under a small lake in the Mackenzie River delta near Inuvik. The strong thawing effect of the lake was borne out by the presence of thawed ground under the lake to a depth of several hundred feet (W. G. Brown *et al.* 1964, Johnston and Brown 1964 and 1965). Further investigations of a similar nature were carried out in April, 1964, in a small isolated lake near Inuvik but away from the delta. Again a thaw basin several hundred feet deep was found beneath the lake.

Permafrost investigations have been carried out in other areas at various times: in 1954 a hole was drilled to a depth of forty feet in a pingo near the Mackenzie River delta revealing the existence of a large body of massive ice within this peculiar landform (Pihlainen *et al.* 1956); in 1955 a permafrost distribution survey was conducted along the Mid-Canada Line; in 1958 an intensive drilling and sampling operation was carried out at Fort Simpson, NWT, on the Mackenzie River (where permafrost is discontinuous), in order to obtain information on the occurrence of permafrost, soil types, and ice content in cleared and uncleared areas (Pihlainen 1961); in the same year, at Norman Wells, two holes were drilled to 200 feet, and in these thermocouple cables were installed to determine the depth and long-term performance of permafrost in the area; and in 1961, thermocouple cables to measure ground temperatures to depths of 200 feet were installed in drill holes at the site of a proposed asbestos mining development in continuous permafrost near Sugluk in northern Quebec.

The Division has assisted the McGill Subarctic Research Laboratory and the Iron Ore Company of Canada in studying the distribution of permafrost in the iron mines at Schefferville, PQ. Here the perennially frozen condition of the ore has given rise to a variety of problems in addition to the usual ones encountered with frozen soil. In 1966 the Division established a field station at Thompson, Manitoba, to study permafrost conditions in northern Manitoba which assume added importance with respect to future developments in this area and in particular the proposed large-scale hydro power developments on the Nelson River. Farther north in the continuous zone, the extensive oil exploration programme and potential production in the Mackenzie River basin and the Arctic Islands is providing impetus to the study of permafrost conditions and related engineering problems.

Techniques for site appraisal to facilitate the investigation of permafrost have been under study for some years. For example, special drilling and sampling techniques have been developed to obtain undisturbed cores of perennially frozen ground and record the soil type and ice segregation. The possibility of utilizing geophysical methods for subsurface exploration in permafrost areas was investigated during the summer of 1958 when field studies were conducted at Norman Wells, Fort Simpson, and Inuvik to evaluate the use of a portable refraction seismograph and an earth resistivity device in determining the depth to permafrost. A detailed study of the factors affecting the measurement of ground temperatures in permafrost areas, with particular emphasis on the use of thermocouples and various types of reading instruments, was carried out to improve the reliability and accuracy of such measurements (G. H. Johnston 1963a, 1966a). A guide to the field description of permafrost was prepared for use by engineers to achieve a simple and uniform description of permafrost in site investigations of potential construction areas (Pihlainen and Johnston 1963).

The large body of permafrost literature now available is under continuous scrutiny and reviews of several aspects have been undertaken. These include reviews of benchmarks, tower construction, and pile foundations in permafrost areas, and the strength characteristics of frozen ground (G. H. Johnston 1966c; Pihlainen 1959). Continual review is being made of the extensive Russian literature on permafrost and selected papers have been translated. There is now an active exchange of permafrost literature between the National Research Council and several agencies in the USSR which are studying permafrost.

The growing interest in permafrost is well indicated by the widespread investigations of other Canadian organizations. Several federal government departments are involved in studying various aspects of permafrost. The Geological Survey of Canada has worked throughout the Canadian North. Their reports on many parts of the North inevitably include many references to permafrost (Brandon

1965, Boyle 1955, 1956, Fyles 1963). The Geographical Branch, also attached to Canada's Department of Energy, Mines and Resources, until it was disbanded in 1968, conducted numerous geomorphological studies in permafrost areas as well as specialized studies of patterned ground and ground ice in the western Arctic (Cook 1958, Ives 1962, Mackay 1962, 1963, Stager 1956). The current interdisciplinary investigation of the Canadian polar continental shelf on the northwest edge of the Arctic Archipelago is similarly relevant.

The Defence Research Board has for many years sponsored investigations of muskeg (peatland) which has important implications for permafrost, and the physical properties of frozen soil. Two branches of the Department of Transport are particularly concerned with permafrost: the Meteorological Branch is studying the climatic aspects of permafrost; and the Construction Branch is developing airfield design and maintenance criteria suitable for permafrost regions. The Department of Public Works, serving as the main federal government construction agency, is very much concerned with the design, construction, and maintenance of many engineering structures in northern Canada.

As well as being involved with the erection of numerous buildings, this department is directing the construction of roads. In Yukon Territory an extensive road network now exists between the sixtieth and sixty-fifth parallels in the discontinuous permafrost zone; the gold mining centre of Yellowknife can now be reached by road. Canadian National Railways are also involved with the design and maintenance of railway lines in permafrost regions: the Hudson Bay Railway in northern Manitoba has been under their jurisdiction for many years and the Great Slave Railway to the south shore of Great Slave Lake, built specially to serve the base metal mines at Pine Point, NWT, passes through the discontinuous permafrost zone (A. V. Johnston 1964).

Farther to the North the Distant Early Warning (DEW) Line, was constructed along the arctic coast, from Greenland to Alaska. Logistics probably constituted the major problem in this gigantic undertaking, since the actual structures used are not large or unduly complex and involve only the careful adoption of accepted foundation methods for permafrost conditions. The wide use of aerial photographs for preliminary reconnaissance and selection of these sites was a design feature of interest.

Numerous building contractors, mining and oil companies, and other concerns, as well as provincial government departments, deal with permafrost in their northern operations. The long-term experiences of the Hudson's Bay Company throughout the North, and the Imperial Oil Company at Norman Wells have already been mentioned. Several mines have been operating for many years and exploration is being carried out by many companies. The movement of equipment during the exploration phase and the provision of services if a mine comes into production are greatly influenced by permafrost conditions. Drilling for oil on the mainland has been underway for a decade and in several places in the Queen Elizabeth Islands. A deep hole drilled on Melville Island by a group of oil companies was instrumented with a temperature-recording thermistor cable to a depth of 2000 feet. The thickest permafrost in North America was subsequently recorded extending to a depth of 1500 feet (Jacobsen 1963).

Most of the agencies whose interests are described above, including the Division of Building Research, are represented on the Permafrost Subcommittee of the National Research Council's Associate Committee on Geotechnical Research. It meets at least once per year to discuss current and proposed investigations, thus keeping all interested organizations informed on developments. In Ottawa in 1962 the Permafrost Subcommittee sponsored the First Canadian Conference on Permafrost which was attended by about 180 delegates from across Canada. A regional conference, attended by the same number of people was held in Edmonton late in 1964 to discuss engineering problems posed by discontinuous permafrost in western Canada.

The Third Canadian Conference on Permafrost in January, 1969 was attended by nearly 400 people from across Canada and the United States. The steadily increasing interest in permafrost and related engineering problems was given sudden impetus by the discovery of tremendous oil reserves on the arctic coastal plain of Alaska in the continuous permafrost zone and an upsurge of oil exploration activities in the Mackenzie River Delta and Arctic Islands in Canada. The realization of the strong influence of permafrost on oil production and pipelining emphasized the appropriateness of the theme of the Conference – permafrost problems related to the mining and oil and gas production industries – and the location in Calgary, the oil capital of Canada. As interest in permafrost continues to grow with increasing economic developments in northern Canada, similar conferences will be convened periodically in the future to provide a forum for discussion of other important aspects of permafrost.

Permafrost research in Canada has come a long way from its modest beginnings. Much information on the scientific and engineering problems associated with permafrost has been gathered but many still await solution, and permafrost research must continue to expand in conjunction with increased economic activities in northern Canada.

B   SETTLEMENT AND TRANSPORTATION ROUTES

The population of Canada's permafrost region is sparse by southern standards. This can be illustrated by comparing the Yukon Territory and Northwest Territories, which comprise most of the permafrost region, with the provinces. The total area of the Territories is 1,511,979 square miles which is 40 per cent of Canada's land area. Their total population is about 43,000 (1966 Census) or 0.03 persons per square mile. Canada's provinces total 2,333,163 square miles with a total population of approximately 20,000,000 or about 9 persons per square mile. Thus, the population density of the Territories is approximately 0.3 per cent of the provinces. A rough estimate of the total population of the permafrost region including people in the northern parts of the provinces would be about 100,000 or 0.5 per cent of the country's population in half of its land area.

More than one-third of the population of the permafrost region is concentrated in about a dozen centres, each of several thousand inhabitants, the largest being Thompson, Manitoba, with a population in 1969 of about 20,000. Virtually all of the remainder live in the many small settlements, each with only a few hundred inhabitants, scattered along the Mackenzie and Yukon River systems and the sea coasts. There are also a few small settlements in the interior of Keewatin District west of Hudson Bay (Woods and Legget 1960).

The Mackenzie River is served by water transport during the summer months and by air throughout the year. A road network has been under development for several years with a stretch of the Mackenzie Highway now extending around the north end of Great Slave Lake to Yellowknife and branch roads currently under construction to Fort Smith and Fort Simpson. Railway links with the south are provided by a line to Waterways, Alberta, where freight is trans-shipped to the Mackenzie River system, and by the Great Slave Railway which extends to the base metal mines at Pine Point and Great Slave Lake at Hay River. Settlements in Yukon Territory are served by the Alaska Highway and branch roads while additional links with the south are provided by coastal shipping to Skagway, Alaska, and by the 110-mile-long White Pass and Yukon Railway to Whitehorse.

East of the Mackenzie River system most of the small settlements – scattered over vast distances – are accessible only by float or ski-equipped aircraft. The existing road networks in northern Saskatchewan and Manitoba are being steadily expanded to reach the majority of settlements in the northern parts of these provinces. Manitoba has the additional links provided by the Hudson Bay Railroad from The Pas to Churchill and a line to Lynn Lake in the northwest. In the permafrost region of northern Ontario, however, there is little settlement. The most northerly land transportation route is the Ontario Northland Railway extending to Moosonee at the southern limit of the permafrost region. East of Hudson Bay, the railway from Sept Iles, on the Gulf of St. Lawrence, northward to the iron mines at Schefferville, PQ, in the centre of Labrador-Ungava, provides the only land communication to the interior. Apart from Schefferville and other mining settlements farther south on branch lines of the main railway, the interior of Labrador-Ungava is virtually uninhabited and accessible only by float and ski-equipped aircraft.

Of a number of small settlements scattered along the coast, those on Hudson Bay and farther east are served by ship from Montreal during the summer, settlements on the arctic coast of Mackenzie and Keewatin districts are served by steamers bringing supplies from Tuktoyaktuk, the trans-shipment point between the Mackenzie River and coastal transport, and ships sometimes bring freight directly from Vancouver through Bering Strait. Between these two sections of coast is an area north of Hudson Bay where access by water transport is difficult because of heavy sea ice. For this reason, there is not an integrated coastal transport system similar to the Northern Sea Route of the USSR. Some of the settlements on the coasts of the Arctic Islands are served by ship during the very short navigation season lasting only a few weeks. The remainder are accessible only by air because of heavy sea ice. Commercial airlines also provide links with flights several times weekly from Montreal to Baffin Island and Resolute, and from Winnipeg to the central Arctic.

Economic developments in the permafrost region have been influenced greatly by the patterns of settlement and transportation. The population is sparse and scattered because of the vastness of the area, inaccessibility of large inland areas and remoteness from southern Canada. The adverse influence of these factors on developments in this region of marginal human activity is aggravated by the presence of permafrost. The already complex natural environment is further complicated by the introduction of human activities. This results in man's being forced to modify his procedures to counteract the problems caused by permafrost. In some parts of the permafrost region these problems are more difficult than in others. Until about 1950 the needs of most settlements were met by small simple buildings. Since that time, as the population increases, larger and more complicated buildings and other engineering structures have been under construction to meet the requirements of centres growing in response to resource development.

4 Buildings

Buildings of many types and sizes exist in Canada's permafrost region ranging from small houses to schools, hospitals, and large heavy-service and mining buildings. It is only within the last ten or fifteen years that large-scale construction has taken place and consideration given to providing permanent stable foundations for superstructures. Previously, most buildings were relatively small, light-weight, low-cost structures which could tolerate some movement. The thawing of the underlying permafrost, particularly in areas of fine-grained soils with high ice contents, reinforced by annual frost heaving in the active layer caused differential settlement of buildings. When this happened, the building was jacked level and wedged with wood, stone, or other available material. It was an annual occurrence and these remedial measures were considered nothing more than regular seasonal maintenance, if in fact they were taken at all.

It was only during the past two decades, when large sums of money were to be invested in buildings that engineering consultants were called in to provide foundation stability. Although northern construction techniques have been improving steadily during this recent period, some problems have arisen. Unexpected soils and permafrost conditions at some locations, inadequate site investigations, and poor recommendations have resulted in difficulties in some buildings.

The present state of building technology makes possible design and foundation construction in permafrost that will perform satisfactorily. Even so each site must be examined carefully both above and below ground to determine proper foundation design. Site investigations vary not only with each location, but even with each segment of each location (Pritchard 1966).

A   FOUNDATION DESIGN

The choice between surface and buried foundations for a particular building will be based on consideration of the soils and permafrost conditions, and the proposed function and life expectancy of the building. The site will have to be prepared for construction and certain precautions should be exercised to prevent foundation failures.

1 *Foundation Types*

All foundations placed on the ground surface are affected by frost action and are therefore frequently used for temporary or low-cost buildings where some movement of the building is not serious enough to interfere with its use. The most commonly used types of surface foundations are gravel pads, mudsills, timber pads, and concrete pads (Pihlainen 1955). Gravel pads are usually three to four feet thick and are used for small buildings (Figure 17). Mudsills are logs or heavy timbers placed about three feet apart under the sills of a building and at right angles to them. Timber pads consist of surface spread footings made of logs or heavy timbers which are placed under the sills or under posts to the sills. Concrete pads are rectangular blocks about eighteen inches square of plain or reinforced concrete which are placed under the sills or under timber or concrete posts to the sills. Because mudsills and timber pads are made of wood, they tend to rot at the ground surface where they are subjected to continual wetting and drying. The average life of such foundations, if not treated with a preservative, is five to ten years. Concrete pads have the advantage of greater durability.

Buried foundations are used for more permanent structures and for those which

FIGURE 17 Gravel pad three to four feet thick at Inuvik in continuous permafrost zone. This type of foundation is used for small buildings which can tolerate some movement.

will impose heavy loads on the foundations. These foundations depend on the perennially frozen ground to carry the load of the building and also to act as an anchor against frost action. Foundations must be buried deeply enough in the permafrost so that there is no possibility of the heat from the building penetrating the frozen ground and thawing it under the foundations. If this is not done, then settlement of the building will occur.

The most commonly used types of buried foundations are timber posts, concrete piers, concrete spread footings, concrete wall footings, timber pads, and piles (Pihlainen 1955). Timber posts are logs or squared timbers which extend down to the permafrost table or are embedded a short distance into the permafrost. They have the disadvantage of being affected by frost action in soil and they rot at the ground surface. Plain or reinforced concrete piers are embedded several feet into the perennially frozen ground so that frost action in the soil will not heave them. Concrete spread footings consist of plain or reinforced concrete pads, embedded in the permafrost, which support concrete posts. Concrete wall footing, also embedded in the permafrost, consists of continuous pads under walls or floors. Another type of foundation is the timber pad consisting of heavy timbers or logs with layers at right angles to one another, the spaces between being filled with sawdust or wood shavings.

In addition to the high cost of materials, the construction of buried foundations is complicated by the difficulty of excavating frozen soil. Methods used may include stripping the surface vegetation to expose the underlying soil to the sun. Thawing amounts to a few inches per day and the thawed material is removed at frequent intervals. Frozen soil can be thawed also with water or it can be removed with explosives or pneumatic hammers. These latter two methods are the quickest and most desirable because they limit the extent of thawing and thus help to attain rapid refreezing, but are the most expensive.

Piles are suitable as foundations for buildings with heavy floor loads or for building sites that are lowlying and poorly drained, although the framing of beams and floors is more difficult and costly than with other types of foundations. Considerable use of pile foundations is being made in Canada's permafrost region, particularly in the Mackenzie River valley because native trees can be used, limited excavation of frozen ground is required, and they can be adapted easily to preserving the permafrost. Piles can support buildings even where it is impossible to maintain the permafrost table at its original high level (Pihlainen 1959; G. H. Johnston 1966c).

Piles are frozen into place in holes in the permafrost. The holes are prepared by steaming or drilling (Figure 18). Steaming penetration rates to produce sufficient thaw for a pile will vary with the type and temperature of frozen soil being thawed. Drilling tends to be more costly than steaming but offers advantages in reducing thermal disturbance of the permafrost. This may be of special value for erection of structures near the southern limit where the temperature of the frozen ground is close to thawing before any construction takes place.

The pile, which may be of wood, precast concrete, or metal, transfers its load to the frozen ground through the adfreezing strength of the soil, end-bearing being too minor a factor for consideration in design (Figure 19). The forces formed by the freezing of all or part of the active layer tend to lift the piles. These forces are particularly significant in fine-grained soils such as silts which are highly frost-susceptible (Figure 20). To overcome this uplift force, the pile should be placed in the permafrost to the depth at which the adfreezing or friction force of the frozen soil plus the load on the pile exceeds the uplift force. A general rule of thumb is that a pile is embedded in the perennially frozen ground to a depth equal to at least twice the thickness of the active layer occurring during the life of the structure. The pile may be anchored in the permafrost by means of shoes or collars, or, occasionally, the pile may be greased or loosely wrapped with tarpaper over that part of its length situated in the active layer in order to prevent adhesion of the frozen soil.

FIGURE 18 Pile steaming and driving in continuous zone at Inuvik. These piles extend about twenty feet into the permafrost. Note gravel pad to protect surface vegetation from construction machinery.

2 *Construction Techniques*

It is preferable to locate buildings on perennially frozen soils that are coarse-grained. Standard construction techniques can then be used because the thawing of these soils may or usually does not have any adverse consequences. They are found frequently in a loose condition such that some settlement will occur on thawing. If a structure must be located on fine-grained soils or organic material, the soil should be preserved in its frozen state for the life-span of the building. The permafrost table may be kept near the ground surface and/or raised either by ventilation or insulation. With ventilation a clear space of at least two feet is provided beneath a well-insulated floor so that cold air can circulate beneath the building and minimize heat flow to the ground. This requires good floor insulation in order to maintain comfortable conditions in the living area above.

A layer of gravel one or two feet thick can be placed on the ground surface prior to construction. This helps to protect the moss cover although the moss becomes compressed and loses its insulation value. The gravel provides some insulation but does not compensate for the moss in its natural state. When the insulation method is used in construction, the building is placed on a pad of non–frost-susceptible material such as gravel (Figure 17). This pad should be several feet thick and should extend beyond the limits of the building and be graded to

FIGURE 19 School at Inuvik, built on piles. Note air space under building to reduce heat flow into ground from building.

drain surface run-off. For temporary structures the pad is usually placed on the ground surface.

The foundation types and construction techniques being discussed are used with the aim of preserving the perennially frozen state of the soil. It is used in the continuous permafrost zone and in the discontinuous zone where the frozen ground is thick and has a temperature several degrees below 32° F. In the southern fringe of the discontinuous zone permafrost is patchy, limited in extent, and close to the thawing point. It may be most suitable to thaw the permafrost prior to construction by stripping the surface vegetation. Undesirable frost-susceptible soils can be excavated and replaced with gravel. Standard construction techniques can be used and the heat from the building will be sufficient to prevent the permafrost from being re-established.

B EXPERIENCE IN NORTHERN CANADA

Most early foundations were simple mud sills of local timbers laid in gravel or sand and levelled with the same material. The sills supported the main beams of

FIGURE 20 Warehouse at Norman Wells, NWT, in northern part of discontinuous permafrost zone. Uneven floor caused by heaving and settlement of the piles due to frost action in the active layer and thawing of ice-laden permafrost.

the building or superstructure. In winter, buildings were banked with sand, sod or snow for protection from the cold and wind. Heat losses through the floor, however, lowered the permafrost table resulting in shifting and heaving of foundations. To avoid this, buildings were raised above ground so that heat losses would dissipate by free circulation of air. Other buildings were placed on thick gravel pads to prevent heat losses from reaching the permafrost.

As buildings in northern Canada progressed from small cabins to the more substantial structures of the trading companies, missions, mining and oil companies, and the federal government, foundations changed from mud sills or gravel pads to piles or piers. These were even combined by placing a gravel pad on the natural cover and raising the structure to provide for passage of air between the underside of the building and the surface of the pad. As buildings grew in size and value the problem of adequate foundations had to be solved. The basic question was whether to accept or to prevent movement of the foundation. Generally the decision was to accept some movement in small, unheated buildings of lesser importance, but to prevent movement in large, heated, permanent buildings. To avoid movement it was learned that the underlying permafrost must remain undisturbed and that precautions should be taken to preserve the moss cover and to prevent large heat transfers from buildings to the ground.

In any practical assessment of building techniques in northern Canada, matters of economics, availability of local materials, and methods and costs of transportation are of overriding importance. These factors have enhanced the use of wood piles in western areas of northern Canada where local timber and river transportation are readily available, whereas concrete and steel are more suitable and commonly used in the sparsely forested or treeless central and eastern regions. In the Mackenzie River region where local timber is available, many building sites occur in areas underlain by fine-grained soils and organic materials which simplify the placing of piles; however, concrete foundations are better suited for sites in the central and eastern regions where heterogeneous coarse-grained soils and bedrock are more prevalent (Pritchard 1966).

1 Yukon Territory

Numerous early buildings suffered damage by permafrost over the years since settlement began. For example, the town of Dawson, which underwent rapid development during the Klondike gold rush at the end of the nineteenth century, is one settlement which has experienced considerable building problems because of permafrost. It is situated in the northern part of the discontinuous permafrost zone, in an area of rolling hills and wide river valleys. The townsite is at the junction of the Yukon and Klondike Rivers and the shape of the settlement is determined by the limits of the river flat. In spite of its situation in the discontinuous zone, perennially frozen ground underlies the entire site to a depth of about 200 feet. The permafrost table is close to the ground surface in most sections except in the south where the Klondike River has deposited sand for a distance of about 600 feet from its present bank (Gutsell 1953).

The effects of permafrost are evident throughout the central part of the town: a park beside the administration building has assumed a roller coaster appearance because thawing of the permafrost has caused as much as fifteen feet of ground subsidence in places; and buildings display severe settlement, tilting, and warping because of thawing of the underlying permafrost, which contains large quantities of ice, and intensive frost action in the active layer (Figure 21).

The hospital, built in 1901, was used for about 15 years and after a period of disuse put back into operation in 1952. There is a basement with plank walls and dirt floor; the foundation consists of log posts, 8 feet long about 1 foot in diameter, resting on wood plates. Movement of the posts by as much as 4 inches because of thaw settlement and frost action require them to be levelled by shimming or cutting off about every second year.

FIGURE 21 Store at Dawson, YT, in northern part of discontinuous permafrost zone. Uneven (differential) settlement is due to thawing of the underlying ice-laden permafrost.

The Territorial Commissioner's residence was erected in 1901 and occupied until 1914. It then remained unused until 1951 when it was opened as a senior citizens' home. Considerable renovation was required because of differential movement through the years. Ten-inch diameter wood posts which, resting on wood sills, support the building, have moved up and down a total of 5 inches and shimming or cutting of them is carried out as required.

The two-storey administration building was constructed in 1900 and has been in continual use since that year. Under the building is a full basement with a wood floor in places and a dirt floor in other sections. The building itself is founded on 10-inch diameter posts resting on mud sills. Some of the posts have had to be shimmed because of thaw settlement.

The public school, which was built in 1901 and occupied continuously, has required maintenance every year because of differential movement. A service garage was built about 1935. It is supported by 15-inch square concrete posts around its perimeter, which rest on concrete pads on the ground surface. Settlements of as much as 3 feet have occurred along the side walls at the centre of the building. The Pearl Harbour Hotel, erected in 1899, has been severely racked by thawing permafrost and water from melting ground ice seeps into the basement. The Alexandra Hotel has settled differentially causing warping and twisting of the floors.

In recent years new mining centres have developed in the Yukon Territory with large heavy buildings erected on permafrost. Clinton Mine of Cassiar Asbestos Corporation Limited is one such settlement located in the northern part of the discontinuous zone 50 miles northwest of Dawson where permafrost is widespread and at least 200 feet thick.

The soils at the townsite are sand, silt, and gravel. The depth to permafrost was about 5 feet and ice contents were fairly high. The mean annual temperature in the permafrost at a depth of 10 feet was observed to be about 28° F to 29° F. It was decided to place all buildings on wood piles with 8-inch diameter tops, drilling 18-inch diameter holes to a depth of 20 feet (Drewe 1969). Specially hard tungsten carbide tipped auger bits were used on a truck-mounted drill after ordinary auger bits proved slow and unsatisfactory in the hard frozen ground. The excavated material was backfilled around the piles for the bottom 13 feet and the top 7 feet filled with gravel. Piles were placed for several buildings in 1966 and construction continued into 1968.

In the plantsite area, 6 miles from the town, the soils are a mixture of relatively dry fine sand and gravel with some layers of silt at one end of the site. The permafrost table varied in depth from 5 to 8 feet; ice contents were low in the sand and gravel and higher in the silt. The mine buildings were set on concrete footings placed in excavations 9 feet deep made by drilling and blasting. The silt area was avoided and the low ice contents of the coarse-grained soils suggested that thaw settlement, if it occurred, would be quite tolerable. There has not been sufficient time since these buildings, and those at the plantsite, were completed, to assess their performance on permafrost. However, it is reasonable to expect satisfactory performance because proper measures were taken in dealing with the local soils and permafrost conditions.

The experience with the early buildings in the Yukon Territory indicates the difficulties imposed by permafrost when little was known about its properties or how to cope with the problems that it caused. New towns, such as Clinton and Anvil, in the central Yukon Territory, which are coming into existence in response to the development of mineral resources, have much larger heavier buildings. Advances in building technology in recent years have improved design and construction techniques of these new buildings; satisfactory performance can be expected if proper attention is paid to the soils and permafrost conditions.

2 *Mackenzie District*

Most of the settlements are located in the Mackenzie River valley. It is underlain by middle Palaeozoic and Mesozoic sediments on which glacial and alluvial materials have been deposited. From Hay River, on the south shore of Great Slave Lake to Tuktoyaktuk on the arctic coast at the mouth of the Mackenzie River, permafrost distribution varies from patchy to continuous.

A survey of buildings in 1950 in this region revealed that many had suffered considerable damage from permafrost. The following descriptions of some of these buildings are taken from the report by Pihlainen (1951). Considerable advances in building technology over the past two decades have resulted in much improved and satisfactory performance of structures erected during this period. The design and construction of some of these recent buildings are described in the following account as typical of the new developments taking place in the Mackenzie District as they are throughout the permafrost region of Canada.

At Hay River the old townsite is located on an island and the soils are fine-grained, ranging from fine sands to clays. Permafrost occurs in scattered islands, its thickness varying from 5 to 50 feet. The depth to the permafrost table ranges from less than 2 feet to more than 10 feet. Temperatures of the permafrost are close to 32° F and ice contents of both perennially and seasonally frozen soils are high. Because of the patchy distribution of the permafrost, it is not known whether the ground is perennially frozen at a particular building site until subsurface investigations are conducted by drilling or test pits. The limited extent of permafrost and the proximity of its temperature to 32° F makes it possible frequently to thaw the permafrost by stripping the surface vegetation – the foundation is designed for engineering properties of the soil in its thawed state – and the heat from the building prevents the permafrost from re-establishing itself.

Here too permafrost has caused damage to a number of buildings (Figure 22), for example in the case of the Royal Canadian Corps of Signals Radio Station which was built on silty fine sand. In the vicinity of the site patches of permafrost were encountered 5 feet below the ground surface. In August, 1947, the excavation of the basement began: a gravel mat about 1 foot thick was laid at the bottom of the excavation; concrete wall footings were 18 inches wide and 12 inches deep; an 8-inch concrete floor and a 9-inch concrete wall were placed for the basement; and no reinforcing steel was used in any of the concrete. The building was completed in December, 1947 (Pihlainen 1951).

One month after completion, cracks appeared in the basement walls and floors. Settlement from thawing of the underlying frozen soils occurred during the spring of 1948 to such an extent that the basement floor practically disintegrated and the cracks on the basement walls ranged in width from hairline to 2 inches. A layer of sand and gravel 1 foot thick was placed on the floor; 7 concrete piers, 1 foot square and 5 feet long, were placed in the middle of the floor where settlement had been most pronounced, and a new 7-inch thick concrete floor was placed. In June,

FIGURE 22 Garage at Hay River, NWT, in southern fringe of discontinuous permafrost zone. Sagging at the centre of the building is caused by thaw settlement of the underlying ice-laden permafrost.

1950, cracks were noticed in this floor. In March, 1948, an attempt was made to repair the large basement wall cracks by excavating the backfill on the walls and patching the cracks with tar and tarpaper. This attempt was apparently not successful because water sceped through the cracks in the spring. The walls were not backfilled then and in 1950 it was possible to look through the cracks to the outside.

Many buildings in Hay River have been built on wood piles driven into place without need for drilling or steaming because the temperatures of the perennially frozen soils is less than 1° F below 32° F. A new development found favour in the early 1960s (Pritchard 1966), when a form of pipe pile being used successfully for foundations was reported to permit use of full basements. Six-inch diameter steel pipes were driven by well-drilling equipment through the permafrost to bedrock at depths of 60 to 75 feet. The pipe lengths were welded together and filled with concrete as they were driven. The economics of this technique had not been studied up to 1963 but its use increased in Hay River which is connected by road to the Alberta oil fields where 6-inch pipe casing and well-drilling equipment were readily available.

A new townsite was selected about 1962 on higher ground on the mainland about five miles south of the old town to accommodate anticipated expansion of economic activities and population accompanying completion of the Great Slave Lake Railway and increased federal government construction of schools and other facilities. This site has better drainage and soil conditions than the old town which is situated on fine-grained soils; the soils at the new site are stony and dense. The distribution of permafrost is similar – scattered islands differing in extent and thickness containing variable quantities of ground ice. Because of these uneven conditions, site investigations are a prime requirement for all construction.

Consideration has been given to adjustable foundations for low-cost housing at the new site. The approved procedure is to lift the building with jacks to the level position and hold it there by shims or extension of the columns as the permafrost thaws and settlement occurs. One problem with this design is that footings may settle differentially or each at different rates and hence place stresses on beams and other components of the building. To achieve satisfactory performance this type of system requires regular maintenance, perhaps for several years, until the ground thermal conditions come to equilibrium.

One building at the new townsite which suffered considerable damage is a hospital constructed about 1967. The soils at the site are stony and very dense and permafrost with ice was known to exist there. The foundation consisted of piles to extend down to firm strata and support the building by end bearing. The piles were not placed to the design depth and settlements of several feet occurred when the permafrost thawed. This example of the proper design for the local permafrost conditions but poor construction practice indicates the care that must be exercised in the discontinuous zone of the permafrost region.

The largest settlement in the Mackenzie Valley is Yellowknife. Here on the north shore of Great Slave Lake, in the discontinuous permafrost zone, permafrost is more widespread than at Hay River and its thickness exceeds 100 feet in some areas. It has caused a number of building failures (such as the Hudson's Bay Company manager's residence), but, generally, sound construction practice has resulted in adequate performance. Pile foundations have been used considerably at Yellowknife, particularly for buildings at Giant Yellowknife Mines Limited. In these cases careful studies were made of ground conditions at the individual building sites and serious difficulties were avoided. Wherever possible, foundations were placed on bedrock, but some buildings had to be constructed on perennially frozen clay. Here special techniques were employed to ensure against severe permafrost damage. At each of these sites, small boxes approximately two feet square by one foot deep were built at the proposed location of every pile. This method was employed possibly to prevent the stony fill from collapsing and falling into the hole steamed for the pile. The sites were backfilled with mine muck fill (waste rock from mining operations) to level them and protect the moss cover from construction equipment. These special construction techniques increased construction costs but the permafrost was protected from thawing (Pihlainen 1951).

The settlements of Fort Providence, Fort Simpson, and Wrigley are located northwest of Great Slave Lake in the discontinuous zone. The distribution of permafrost varies from scattered islands 25 to 50 feet thick at the first two settlements (Pihlainen 1961) to generally widespread in the vicinity of Wrigley. The soils at these settlements are fine-grained with high ice contents, and numerous old buildings with simple foundations erected years ago have suffered damage from permafrost and thaw settlement and frost heaving. Little knowledge or experience was available on permafrost conditions or methods of coping with construction problems caused by them. However, the situation has changed considerably in recent years with the increase of knowledge and experience in coping with the engineering problems caused by permafrost. New buildings such as houses, schools, hospitals, and heating and power-generating plants have been constructed and are performing satisfactorily even in the adverse soil conditions which prevail in the Mackenzie River valley.

Both pile and poured-concrete foundations have been used at Fort Simpson in recent years (Pritchard 1966). Construction of a school and children's hostels were planned during erection of similar buildings at Inuvik about 1959. It was anticipated that the same type of wood pile foundations would be used. Site investigations, however, showed some permafrost-free areas. The original mission hospital was built on a log mat in 1913; in 1952 an addition was built on a concrete foundation with full basement. No settlement occurred in either part of the building. Several other concrete buildings in the area had remained trouble-free, so the buildings were designed using concrete piers, spread footings and partial basements.

Soil samples taken from the basement areas after excavation to the full depth showed ice lenses at depths from 6 to 9 feet and permafrost in half the school excavation, while the other hostel had been excavated in an area free of permafrost. The samples revealed that in some areas the soil contained ice lenses and would not sustain the calculated foundation loads. The area relatively free of permafrost and ice lenses had been cleared of brush and trees some years before, as was the case in all other areas where full basements had been built without difficulty. Areas where permafrost occurred had been cleared just prior to excavation (Pihlainen 1961). It was decided to build the buildings as planned but to redesign the foundations. Further test borings indicated a hard resistant layer 20 feet below the basements and piles were driven to this depth. Some 2,400 wood piles, 25 feet long were driven in clusters; a continuous reinforced pile cap served as the footing for the foundation walls.

Norman Wells, 90 miles south of the Arctic Circle, is located in the northern part of the discontinuous permafrost zone. The annual depth of thaw in undisturbed areas varies from less than 2 feet under moss cover to about 10 feet in areas having no moss. The permafrost is about 200 feet thick. The soil is predominantly silt with some thin layers of gravel and clay overlying shale strata encountered at a depth of approximately 40 feet (Pihlainen 1951).

Before the Second World War, Norman Wells was a small settlement consisting of buildings with simple foundations. A small oil refinery, about 1,000 barrels per day output, was in production supplying fuel mainly to the mining operations at Port Radium on Great Bear Lake and small quantities to the small settlements scattered along the Mackenzie River. With wartime expansion of the oil field production and construction of the Canol pipeline (see p. 135), the settlement enlarged rapidly and some large heavy buildings were erected. Because of the presence of permafrost, in some cases the ventilation method of construction was used, whereby an air space was provided between the floor of the buildings and the ground surface to dissipate heat losses from the buildings. Wood and steel piles were used for the foundations of such buildings as the refinery, the heat and power-generating plant, and several warehouses.

The first piles were used in 1943, and because of high shipping costs on imported timber, native spruce was used which ranged from 7 to 10 inches in diameter at the butt, some having asphalt-treated collars to help prevent frost heaving. Piles driven to depths of 12 to 15 feet, with or without collars, had not shown evi-

dence of heaving when they were examined in 1948. In 1950 there was some evidence of pile heaving which appeared to indicate that the permafrost table was
being lowered under some buildings (Pihlainen 1951). (By the late 1960s the
permafrost table has receded to depths of 20 feet and more under some buildings
which have existed since the Second World War.)

Availability of scrap pipe from the oil field operation has led Imperial Oil to
use only steel piles since 1947. These are more easily handled and driven than
wood piles, but because steel has a much higher thermal conductivity than wood
there was speculation on the effect of increased heat flow in the steel piles. Con-
troversy exists over the relative degree of heat conductivity in wood and metal
with respect to the resulting influence on soil strength, but definite answers are
not known.

Norman Wells has a central heating system. Steam from a boilerhouse is carried
to the buildings through pipes in above-ground utilidors – continuous insulated
enclosures or boxes constructed of concrete, wood, or metal; the pipes are distributed under the floor to the various rooms. The skirting board around the
buildings, whose function it is to enclose the air space under a building between
the floor and the ground surface, is used on buildings placed on piles where there
is an air space of several feet, and, along with the heat from the steam pipes, not
only helps to keep the floors warm but also has a marked effect on the thermal
regime of the underlying perennially frozen ground. Temperatures immediately
under the heated buildings are very warm during the winter. In a few years this
heat thawed the permafrost sufficiently to cause sagging of some of the buildings –
one group of six buildings on 4-inch diameter steel piles all settled (in the middle
of the floor) because of thawing of the permafrost. In addition to this, frost heaving
of perimeter piles produced the net effect of leaving the walls an average of 2
inches higher than the middle of the floor. In the fire hall, the utilidor enters the
east wall and the building has settled 4 inches here. The post office canteen building has the utilidor along the side of the building entering the west wall. Here the
southwest corner of the building has settled about 2½ inches. One living quarters
has had to be jacked and blocked 8 inches in places. On the south side of the
laundry building, where heavy machinery is installed, settlements up to 10 inches
have occurred (Pihlainen 1951).

The refinery and building housing steam boilers for heating and diesel units for
generating electricity (boilerhouse) at Norman Wells have both suffered some
damage because of thawing of the permafrost. In these buildings the foundations
supporting the floor and machinery are separate. In the refinery, enlarged to a
capacity of about 2,000 barrels per day after the Second World War, the pump
bases are on 4-inch diameter steel piles driven to depths of 14 to 16 feet and the
floor is on 4-inch diameter piles. There is an airspace of several feet beneath the

building. The floor has settled differentially as much as 2 inches but this has not affected the operation of the pumps. The boiler house has locomotive-type boilers, pumps, and generators. Here, as in the refinery, the boiler foundations are separate from the floor system being on 7-inch diameter steel piles. Considerable settlement occurred mainly because boiler waste water was allowed to collect around the building causing thawing of the permafrost and settlement of the piles supporting the building and the boiler foundations. In 1950, from April to September alone, the perimeter of the building settled 6 inches and the floor 5 inches (Figure 23).

Construction of a new boilerhouse began late in 1950. Steel pipe piles with fins at the ground surface to dissipate downward heat flow were installed to bedrock at a depth of 20 to 25 feet. No appreciable foundation movements have occurred

FIGURE 23   Thawing and differential settlement of the under-lying ice-laden permafrost has caused breakage of the concrete floor slab in an abandoned boilerhouse at Norman Wells.

and the building has been performing satisfactorily. New housing for the power generators was erected and began operating in 1969. The building rests on a gravel pad several feet thick ventilated with 12-inch diameter ducts. Piles were installed to bedrock to support the turbine blocks and around the building perimeter.

A new service building was constructed in the early 1960s incorporating offices, machine shop, mechanical repair shop, warehouse and storage facilities. It is partly one storey, partly two, approximately 100 feet square, and metal clad. The foundation consists of steel pipe piles resting on bedrock at a depth of 20 to 25 feet. There is a gravel pad several feet thick under the building and an airspace of 3 feet – one foot around the perimeter. This building has performed satisfactorily since it was erected and no permafrost problems have been reported.

Most of the oil tanks at Norman Wells were constructed shortly after the Second World War. Site preparation consisted only of clearing the surface vegetation with a bulldozer and laying one or two feet of dirty gravel before the tanks were erected. The plates comprising the tanks were bolted, not welded. Many of the tanks were heated causing thawing of the underlying permafrost. The resulting differential settlement induced stresses and strains, especially critical in bolted tanks, which caused the plates in a few tanks to buckle and the oil to leak out (Figure 24). Apart from some oil leakage no serious difficulties were encountered in these relatively small tanks. Some difficulties with thawing permafrost were encountered in a few large tanks that were also erected.

Oil storage in permafrost regions presents difficulties in tanks depending on the type of product and variations through the year. Design criteria are difficult to establish because crude oil, for example, has to be heated to a temperature exceeding 100° F which imposes considerable heat loads on the underlying permafrost.

In recent years, the federal government has constructed a number of buildings on wood piles at Norman Wells. Some of these have suffered damage similar to the older buildings, because of thawing of the underlying permafrost and frost heaving in the active layer.

Between Norman Wells and the Mackenzie River delta lie the three small settlements of Fort Good Hope, Arctic Red River, and Fort McPherson. The soils, ranging from fine sand to clay alluvium with patches of gravel and organic material, have high ice contents and a surface cover of moss. The permafrost is continuous and lies within a few inches to a few feet of the surface. The situation is similar to that described in the area between Great Slave Lake and Norman Wells in that the older buildings are small and light-weight, constructed on mud sills resting on the ground surface. Thawing of the underlying permafrost and frost action in the active layer beneath many buildings has caused damage due to the resulting differential movement. In these settlements, as in others located on permafrost, common features of such buildings are wall panels pulled slightly away from the floor, undulations in the floor, cracks in masonry and concrete, porches warped and pulled away from the house, and the sticking of windows and doors.

Several large buildings were erected at Fort McPherson during the period from July 1956 to December 1958 (Harding 1963). Bedrock at this site is a hard grey black shale overlain by varying thicknesses of weathered rotten shale. The maximum depth to bedrock at the project site was sixteen feet. Fort McPherson is located in the continuous permafrost zone; large quantities of ice occur in the weathered shale and in fact comprise generally more than half the total volume of overburden.

A hostel for a hundred pupils, classroom addition to existing school, teacherage, generator building, warehouse, and walk-in freezer were constructed on concrete piers. A garage was built on piles. An icehouse was placed on mud sills on a shale pad on the undisturbed ground. An oil storage tank was also placed on a shale pad six feet high. Piles were not used for the hostel because it appeared that there was not sufficient overburden on the shale to place the piles.

FIGURE 24 Tank at Norman Wells holding crude oil has settled because of thawing of the underlying ice-laden permafrost. The plates have buckled and oil is leaking. There is also danger of the tank's being punctured by the catwalk on the right.

Some difficulties with permafrost arose during construction. About 100 excavations, each approximately 4 feet by 4 feet by 11 feet had to be made in the permafrost for the concrete piers. Early work had to be done by pick and shovel until air compressors and jackhammers arrived at the site. The gravel for the concrete piers, which was hauled 40 miles by boat from a beach, was late in arriving and the permafrost began thawing in the excavations. The large ice layers mentioned above melted and massive soil slumping occurred. It became obvious that large quantities of backfill were going to be required, more than could be obtained from the pier excavations because of the large volume of ice. Considerable demands were thus imposed on the limited available equipment.

The generator building was a special case because the pier spacing was too close for individual hole excavation. One large pit was excavated covering an area 68 feet by 45 feet and was 16 feet in depth down to the shale. The area was stripped of its moss cover and the frozen material beneath was broken by jackhammers and the pieces removed by dragline. Severe thawing occurred on the walls of the pit, particularly that facing south. Water from melting ground ice was removed by continuous pumping. The total excavation, forming of the piers, beams, and floor slab took exactly one month from mid-August to mid-September in 1956.

Although some difficulties were encountered with permafrost during construction as described above, no problems have been reported since these buildings were completed and they have apparently been performing satisfactorily.

Aklavik is a large fur-trapping settlement in the northern part of the Mackenzie River system and is situated on the Peel Channel of the Mackenzie Delta. The soils are deltaic fine sands, silts, and clays, with ice content up to six times the volume of soil. Here permafrost is continuous and lies near the ground surface. Disturbance of the ground thermal regime by construction activity causes thawing of the upper portion of the permafrost.

The difficulties experienced by the two buildings mentioned subsequently are typical of those placed on surface foundations at Aklavik. The Anglican bishop's residence, on timber sills settled slightly causing jamming of some of the doors (Pihlainen 1951). The Aklavik Power Company powerhouse engines rested on timber cribs which settled about one-half inch, a critical amount for this type of structure. New larger generators were installed in 1953. They were set on six-foot-thick poured concrete blocks placed in pits excavated through the three-foot-thick active layer and about three feet into the permafrost. The bottom and sides of the pits were lined with rows of logs and sawdust to protect the permafrost from thawing. Upon examination a year later no movement of the generator set had occurred.

A number of buildings at Aklavik have buried foundations, many of which have been damaged by thawing of the permafrost. The Royal Canadian Mounted Police single men's quarters, built on timber posts, showed signs of heaving and the centre of the building is slightly lower than its perimeter. Two private residences have full basements of the conventional type; the walls and floors are concrete and the walls are the supporting foundation of the buildings. The basements have 10-inch-thick concrete walls and 4- to 6-inch-thick concrete floors. No special precautions were taken against the problems of permafrost in their design and construction and both basements showed signs of failure. One basement has severe cracks in the concrete floor and moisture has leaked in at the junction of the wall and floor; the other has badly cracked walls owing to settlement caused by thawing of the permafrost.

By 1954 the permanent population of the town numbered about 400, rising to about 1,500 several times each year when people moved in from the surrounding trapping areas. As the population increased, Aklavik became increasingly important as a federal government administrative centre in the western arctic (Robertson 1955a). In view of its increasing responsibilities in this part of northern Canada, the government had to decide whether to enlarge its facilities at Aklavik or develop another site in the same general area.

Several aspects of Aklavik's site made it unsuited for large-scale expansion (Merrill et al. 1960). The deltaic soils are fine-grained and in their perennially frozen condition they contain large quantities of ice. If the town had been enlarged with additional clearing of the vegetative cover and more heated buildings installed without suitable foundations, further thawing of the permafrost and accompanying ground subsidence would have occurred. Drainage had always been a particularly difficult problem because of the flatness of the town site and imperviousness of the underlying permafrost. There was the possibility every year of flooding when the river reached its highest level, during spring breakup, and the town had been flooded several times in the past. Even when the river was not in flood, it was eroding the bank and the edge of the town site was being washed away. However, sufficient quantities of gravel to build roads and an airstrip were not available locally but would have had to be hauled many miles over difficult terrain. Another adverse factor was that the size of the site, being restricted to a small space between the river and boggy areas, left no practical way for the town to expand.

Because of these factors the federal government decided to search for a new town site. Consequently, during the summer of 1954, a site survey team examined several areas on both sides of the Mackenzie Delta, at about the same latitude as Aklavik, which could serve the same population (Brown 1956). The present site of Inuvik, 35 miles east of Aklavik, is on high ground adjacent to the navigable East Channel of the delta. The soils are coarse-grained, not susceptible to frost heaving, and large quantities of gravel for building foundations and roads are available. The site provided room for a settlement considerably larger than Aklavik with all buildings placed above the river's flood level, secure from river bank erosion. Sufficient land was found for an airstrip which could be linked to the town by road. At the time of the site selection survey consideration was given to moving all the major existing buildings at Aklavik to Inuvik but this scheme was abandoned as not feasible.

Construction at Inuvik began in 1954 and is still continuing. All major buildings, including serviced housing and warehousing are on piles with air spaces of several feet (Figure 19). Two large oil storage tanks are built on concrete slabs resting on piles with air spaces. These buildings have performed quite satisfactorily and only a few piles of the more than 14,000 installed have shown any significant movement. Movement of vehicles and mobile construction equipment was closely supervised to prevent destruction of the natural moss cover (Pritchard 1962). Contractors drove over prepared gravel routes and stored building materials on prepared gravel pads. Small, low-cost buildings – mainly unserviced housing – are on mud sills and air space on gravel pads. These small buildings have also performed satisfactorily. Exceptions are the Roman Catholic church and rectory which have basements.

Roads were built using local gravel material and an airstrip, connected to the town by an 8-mile road, was constructed of local gravel and crushed quarried limestone; services were placed in utilidors supported above ground on piles. Permafrost is several hundred feet thick here (as it is at Aklavik) and lies within a few feet of the surface. However, careful site survey at individual building, utilidor, road, and airstrip locations has indicated that the soil conditions are better than at Aklavik. Construction costs have been higher than they would be for a similar town where there is no permafrost, nevertheless, the cost of expanding Aklavik to the present size of Inuvik with the same large buildings would have been much higher.

On the East Channel of the Mackenzie River delta, about fifty miles northeast of Aklavik, is Reindeer Depot, which was the centre for the reindeer herding industry in the tundra to the north and northeast. In the small building area at the foot of the Caribou Hills the soils are alluvial silts with high ice content and permafrost is continuous and lies near the ground surface. Several buildings have been damaged by permafrost. One is the Hudson's Bay Company store and dwelling, in a building set on concrete posts resting on concrete pads 2 feet square and 9 inches deep placed on the undisturbed ground surface, with an insulated double floor, and a 3-foot airspace skirted with loose boarding. Frost heaving causes gaps as large as 2 inches between partitions and walls to appear during the winter. In summer the building settles and the gaps disappear.

In 1950 a survey was conducted of buildings on surface and buried foundations in settlements on the Mackenzie River system (Pihlainen 1951). The purpose was to obtain an idea of the performance of various foundation types on permafrost. The results, presented in Table 2, disclosed that nearly two-thirds of the building foundations in existence at that time which were investigated had suffered some damage because of permafrost thawing and frost action.

TABLE 2
Foundation Failures Caused by Permafrost on the Mackenzie River System (Pihlainen 1951)

Type of foundation	Number of structures investigated	No apparent failure	Apparent major failure due to:	
			Frost heaving	Settling
Mudsills	76	30	30	16
Timber pads	19	5	5	9
Concrete pads	6	1	4	1
Timber posts	15	7	8	0
Short concrete piers	7	2	5	0
Concrete wall footings	23	10	1	12
Piles				
wood	10	4	3	3
steel pipe	25*	9	4	12
concrete	1	1	0	0
TOTAL	182	69	60	53

*All at Norman Wells

This cursory survey of buildings in the settlements on the Mackenzie River system pointed up the problems encountered with permafrost up to that time. It was perhaps unfortunate that this area, one of the avenues of early development in Northern Canada, is the site of predominantly fine-grained soil with high ice content. These difficult conditions caused many problems in the construction and maintenance of even small buildings. Considerable advances in technology and increasing appreciation of the problems have taken place since 1950 with the result that large, heavy, complex buildings are being erected and performing satisfactorily in these adverse soils and permafrost conditions. Today, as in the past, careful site investigations and foundation design to suit local conditions are absolutely essential to avoid trouble.

3 *Central and Eastern Regions*

Many of the settlements east of the Mackenzie River are populated mainly by Eskimos and/or Indians living in small cabins. The number of white people is small; there are few larger buildings except in towns with relatively large white populations. These include Uranium City in northwestern Saskatchewan, Thompson, Lynn Lake, Gillam and Churchill in northern Manitoba, Schefferville, Wabush, Labrador City, Churchill Falls and Goose Bay in Labrador-Ungava and Frobisher Bay in southeast Baffin Island. Building difficulties caused by permafrost have been encountered at some of these settlements but careful site investigations and construction techniques have helped to ease the problems.

Uranium City is a mining community situated in the northwest corner of Saskatchewan, on the north shore of Lake Athabaska, in the discontinuous permafrost zone. A careful site survey was conducted and a townsite was found which has no permafrost. The soils are sand and gravel grading into almost uniform sand in the lower portion of the site. Natural drainage was blocked by a rock ridge and the area was covered by several feet of peat having a high ground water table. Much of this peat has been removed where buildings have been erected and little frost heaving has occurred in the coarse-grained soils. Within a few miles of the townsite perennially frozen silts containing large quantities of ice were encountered at a road location, causing high construction and maintenance costs. If there had not been a careful site investigation prior to construction, the town might have been located in the frozen silt area resulting in numerous building failures.

The mining town of Lynn Lake in northwestern Manitoba is another illustration of the value of careful site investigations in permafrost regions. This town is located in the northern part of the discontinuous zone where most of the ground is underlain by permafrost. Although the soils are predominantly silt clays with high ice content, a site survey revealed the existence of sand of glacial outwash or deltaic origin near the orebodies. Permafrost was encountered in the sand but it receded when the surface vegetation was removed. Seasonal freezing and thawing have caused little heaving and settling of building foundations in this coarse-grained non-frost-susceptible soil. One exception is the recently constructed railway station described in the section on the Lynn Lake Railway, p. 130.

Thompson, located in north central Manitoba in the discontinous permafrost zone, is the site of a large, recently developed nickel mining and smelting operation. Permafrost occurs in scattered islands varying in extent from a few tens of feet to several acres, and in thickness from about 3 feet to 40 feet or greater, averaging between 8 and 15 feet (Johnston *et al.* 1963). The permafrost table is generally encountered anywhere from 1½ feet to 7 feet below the ground surface. Much ice, primarily in the form of horizontal lenses up to 8 inches thick (the average thickness being less than 1 inch), is found throughout the frozen lacustrine silts and clays underlying the area. Temperatures in the permafrost vary from about 31° F to 32° F.

Construction problems have arisen mainly because of the relatively unpredictable distribution of the permafrost islands, the large quantities of ice contained in the soils, and the near thawing condition of the frozen ground. A minor disturbance or change in local natural conditions, such as removal of the insulating moss cover or erection of a building, causes significant thawing and results in large ground settlements.

The distribution of permafrost is so erratic that there were a number of early buildings with permafrost under part of the structure and unfrozen ground under the remainder. Some of these and other buildings, built entirely on perennially frozen ground, suffered severe damage from thaw settlement (Figure 25). As a result of these difficulties, local regulations were enforced requiring borings at each proposed building site in order to ascertain the absence or presence of permafrost. In the former case, the site could be used immediately; in the latter the surface vegetation was removed and the lot left vacant for several years until the permafrost disappeared. In those parts of the town where permafrost was found there were streets with only one or two houses for a few years until the permafrost had thawed in the other lots.

The town of Churchill is on the west shore of Hudson Bay, in northeastern Manitoba, in the continuous permafrost zone. The soils are a mixture of fine-

FIGURE 25 Basement damage of house built at Thompson, Man., in discontinuous permafrost zone. Differential thaw settlement of underlying ice-laden permafrost has caused basement floor and walls to sag and crack.

grained silts and clays and coarse-grained sands and gravels with bedrock out-crops. Permafrost occurs everywhere and buildings must be located either on the coarse-grained soils or special precautions must be taken if fine-grained soils underlie the foundation. Detailed site investigations are required to locate the best building sites. One example at Churchill of a building foundation designed specially for perennially frozen fine-grained soils was constructed by the Royal Canadian Navy (Dickens and Gray 1960). The soil at the site was a till with high ice content and here the foundation consists of concrete piers 18 inches square designed to raise the structure off the ground sufficiently to allow circulation of air between the underside of the first floor and the ground surface, and thus reduce heat transfer to the soil. The piers are carried on concrete pedestals 2 feet square at the top, 3 feet square at the base, and 6 feet to 7 feet 6 inches high, the sides of which are sloped to reduce the effect of frost-heaving forces. The building load of each pedestal is transferred to the soil through a 4 feet square by 18 inches deep concrete spread footing which rests on the permafrost table. A gravel mat 4 feet thick, covered with an insulating layer of moss 1½ feet thick, extends over the pedestals and footings, and protects the permafrost from thawing.

Construction of the foundations was begun in July, 1948, and completed in August, 1950. A special construction schedule, described here, which was established for the field crew played an important role in success of this building. First, drainage ditches were provided to remove accumulated surface water followed by the excavations for the spread footings. The concrete footings and pedestals were placed on tamped gravel fill. Gravel and moss were stockpiled in the spring of 1949 as fill and insulation for the foundation area. The site was kept clear of snow during the winter of 1948–9 to allow the foundations to freeze in. Before the ground thawed in the spring of 1949, four to five feet of gravel fill was placed over the building area and covered with an insulating layer of moss. Finally, the concrete piers and beams were placed to complete the substructure. Apart from some drainage difficulties at the beginning of construction, the schedule was followed and the job completed satisfactorily.

An examination of the building in 1959 revealed that this foundation, designed to keep the permafrost up to the bottom of the footings for support, and the special construction procedure has resulted in satisfactory performance. It is recognized, however, that this foundation is more complicated and costly than would have been required for a similar building in an area where there is no permafrost.

The largest town in the eastern Arctic is Frobisher Bay on the east coast of Baffin Island. In the past few years it has been the site of considerable development because of its role as an administrative centre for the surrounding area, and as a fuelling stop, for a short span, on trans-polar air routes from western North America to Europe. The soils are predominantly fine-grained with high (ice) moisture content, and the permafrost is continuous. Even before the recent de-velopments mentioned above, a number of buildings had been erected all of which experienced damage because of thawing of the underlying permafrost. These buildings settled as much as three feet and, in fact, the foundation of a diesel power plant settled so much it was feared that it would have to be abandoned. It was essential to keep the plant in operation until the completion of a permanent plant under construction. To do this, refrigerant was circulated in pipes in trenches beside the foundation until the ground refroze and settlement ceased (Spofford 1949). Although this eased the problem, the cost was high and the technique so uneconomical that it could not be considered standard.

When the new town was first proposed in the late 1950s and the most probable site selected, extensive drilling was undertaken to select sites for specific buildings and the types of foundations best suited to the sites. The results of these tests, the varying depth of overburden and the possibility of substantial downslope soil movement led to the decision to build on solid rock (Pritchard 1966).

The main part of the town was to be built on a plateau formed by blasting solid rock from the top of a hill. The hospital, power plant, and water treatment plant were to be built on a rocky hillside immediately north of the townsite. Rock cores indicated granite with some fissures and weathering and little likelihood of ice lenses. Concrete foundations were placed down to bedrock and permafrost did not present any problems. However, considerable seepage occurred in the over-burden during construction and special measures such as deepening sumps and diverting surface drainage were employed.

At Fort Chimo, in the northern part of the discontinuous zone, permafrost is widespread but there are areas free of permafrost mainly where the soil consists of well-drained sand. Construction of a school on concrete footings and a gravel pad was to begin in the summer of 1963 and be completed in 1964 (Pritchard 1966). The active layer at the site varied from five to seven feet and it was agreed that the permafrost must be preserved. The site was to be levelled by cut and fill, and the pad, consisting of alternate layers of gravel and peat, placed and compacted as quickly as possible. It was felt that the peat would help to insulate and preserve the underlying permafrost. This design appears to be performing satisfactorily.

The diversity of soils and permafrost conditions in the central and eastern parts of Canada's permafrost region have resulted in the design and construction of a variety of building foundations. New mining communities have developed and older settlements have been expanded with the construction of new houses, schools, hospitals, and other buildings. Although there have been some problems, and even failures, particularly in early construction, many buildings have been erected in recent years and are performing satisfactorily.

C CONCLUSION

A study of building foundations in Canada's permafrost region indicates that it is difficult to generalize or always predict with certainty the most suitable design and construction techniques. Many buildings have been constructed successfully and there have been many failures, particularly in the early days of settlement. Anything can be built, in any soils and permafrost conditions, provided the conditions are investigated thoroughly and proper design precautions are taken.

In dealing with permafrost it is known that any disturbance of it will cause changes but the degree of alteration is often difficult to assess. How much change takes place in the permafrost when large buildings, several hundred feet in length, are constructed? How much does the permafrost recede along the south wall and advance along the north wall? When some action raises the permafrost table so as to dam an established water course, where will the new flow appear, and will it damage the foundations of some other building? What happens when piles are dependent on refreezing and it fails to occur? (Pritchard 1966)

It is popularly assumed that solid rock presents no problems. Shale that was believed to be solid and hard has turned out to be riddled with ice-filled fissures. Seemingly solid granite has broken down into weathered stone and coarse sand when the ice-cementing mineral particles have melted. It is not safe to assume that rock in permafrost condition will provide a trouble-free bearing for foundations. Design of foundations must continue to be based on the best available site information coupled with the knowledge that designs will probably be altered to suit site conditions as they are disclosed.

The problems of designing foundations in areas are solvable and many have been overcome. There is the necessity of realizing that permafrost exists in a variety of forms, and of understanding the particular variety to be dealt with at a given location. This can be clarified only by thorough site investigations prior to designing the foundation and by studying the performance of the many types of foundations built in various types of soils and permafrost conditions in recent years. There is a requirement for documenting case histories of building successes and failures to assess the entire situation. This will contribute to improved building technology and facilitate economic developments in the permafrost region.

5 Services

The provision of services in towns and villages in northern Canada is greatly complicated by permafrost. Because of high capital and maintenance costs, due in large part to the presence of permafrost, only a few of the larger settlements are equipped with water distribution and sewage disposal systems similar to those in southern Canada. The provision of electricity to larger settlements is also affected in so far as power dams and related water-retaining structures have to be specially designed for permafrost conditions. Although permafrost is a handicap to the provision of services, it has been utilized (for example, as natural cold storage of perishable foodstuffs).

A WATER SUPPLY AND SEWAGE DISPOSAL

1 *Water Supply*

The existence of a water supply in permafrost regions is influenced greatly by low air temperatures and the perennially frozen state of the ground. Supplies of water readily available for use throughout the year are scarce and an adequate supply suitable for domestic and industrial use is an important factor governing the location and development of permanent communities and industries. Accounts of water supply sources and distribution methods in permafrost regions have been written by Alter (1950), Dickens (1959), Legget and Dickens (1959), and Yates and Stanley (1966). The material presented in the two following sections has been obtained mainly from these sources.

a / Sources of supply
There are a number of sources of water for domestic or industrial purposes: surface water comes from lakes and streams, groundwater from above, within or beneath the permafrost; other sources include rainwater and melted ice and snow.

Although large areas of the permafrost region have myriads of small lakes, ponds, and swampy areas which may provide water during the warm season, these are relatively shallow and are completely frozen in the winter. Bodies of water must be more than eight to ten feet deep before they can be expected to provide water throughout the year. Ice cover rarely exceeds eight feet but the storage space in a lake is much reduced when a thick ice cover exists even if some water remains below. The freezing action tends to concentrate the mineral and organic content of the lake in the unfrozen water below the ice, making it unsuitable for use. Many northern lakes would probably not provide adequate storage for a community of more than a few people.

Lakes which are fed by underground springs or receive an appreciable amount of shallow subsurface drainage may provide an adequate water supply, but many lakes are deceptive because they are principally the result of retarded drainage through permafrost rather than evidence of a large and continuous source of water.

Apart from the few major rivers which are large enough to maintain an appreciable flow throughout the year, most streams are small and freeze to the bottom in winter. Some stretches of larger rivers may also freeze to the bottom. The formation of frazil (i.e., formed by the freezing of turbulent water and resembling slush) and anchor ice (i.e. formed on the bottom of lakes and rivers) complicates the utilization of river water. The clogging of intake lines with these types of ice can be controlled with steam lines placed in intake structures.

The three types of ground water which are possible sources of supply in permafrost areas are suprapermafrost water (water occurring between the ground surface and the permafrost table), intrapermafrost water (water occurring in thawed areas in the permafrost), and subpermafrost water (water occurring beneath the permafrost).

Suprapermafrost water sources are irregular and may disappear before the end of winter where the seasonal frost extends down to the permafrost table as it does in the continuous zone. These water sources are generally poor producers and are not dependable for large requirements. They are most common in the discontinuous permafrost zone where permafrost lies at some depth below the maximum penetration of seasonal frost. The water may also be contaminated because it is rarely deeper than ten to twenty feet and may mix with water from that zone in the subsoil where cesspools are located. Another problem is created by the formation of a thaw basin in the ground under a heated building into which the suprapermafrost water will drain – if this happens, it is no longer available as a water source in the vicinity of the building.

Intrapermafrost water is rarely found except in the discontinuous permafrost zone. It may also occur further north in mountains or foothills where subpermafrost water is forced into the perennially frozen ground by hydrostatic pressure or it may occur in a fault zone or talik layer of unfrozen ground between the seasonally frozen ground and the permafrost, applying also to an unfrozen layer within the permafrost as well as to the unfrozen ground beneath the permafrost. It is not a stable supply and may be exhausted or may come through the frozen ground and appear as suprapermafrost water. It can be procured by using drilled, or thawed and jetted, wells. This type of supply can be likened to water supply in fissured limestone. Such supplies differ greatly in quality and quantity .

Subpermafrost water supply is the most promising of continuous water supplies in permafrost regions but it is difficult to locate, costly to develop, and often highly mineralized. Permafrost may be hundreds of feet thick and pervious strata below this may not contain sufficient water.

Drilling through permafrost is difficult and wells must operate almost continuously to prevent freezing of the water in the pipes, yet at the same time there is the danger of well freezing because of excessive pumping. Well casings must be anchored firmly in the permafrost to prevent seasonal frost from crushing, disjointing, or otherwise destroying the casing. Large diameter casings and continuous moderate pumping are helpful in preventing freezing of the well in the permafrost because the tendency for supercooling of the water and frazil ice formation is reduced. A particular advantage of well water is its low turbidity which eliminates the need for large settling basins generally required by highly turbid surface sources. The dynamiting of subpermafrost and intrapermafrost wells to increase the yield is dangerous and may result in possible hazards of water contamination similar to the hazards associated with blasting wells in limestone.

Rainfall is low throughout much of the permafrost region, which makes cisterns an unreliable source of water supply. In summer a small amount of water can be obtained from shallow wells dug into the permafrost in which water from melting ground ice accumulates. The principle water source for the typical small community in the permafrost region is melted ice and snow. Ice is cut from fresh water lakes or rivers in the fall when it is about one foot thick and is stored in a permafrost cellar or an ice house, or simply left on the ground surface. Special melting tanks equipped with steam coils are used in some larger buildings but in most homes ice is put in a container in the kitchen to melt. High costs of fuel and labour needed to produce an adequate supply of water by melting make this method impractical for obtaining the large quantities required for an entire community. During the summer, water for household use is usually obtained from fresh water lakes or rivers by bucket.

Water storage reservoirs are another possible source in northern settlements being considered by the Department of Indian Affairs and Northern Development, and others. A number of serious engineering problems arise, however, in the construction and operation of these structures which have to be resolved to achieve satisfactory performance. The reservoir has to be sufficiently deep so that in winter enough water is available beneath the ice cover to satisfy the requirements of the settlement. The main problem is that providing a reservoir is equivalent to placing a lake on the permafrost which will cause thawing to occur. The actual construction of the reservoir requires the excavation of frozen ground which is in itself a difficult task. Careful site selection is required to choose a suitable location. Pervious strata should be avoided if possible because water will seep from the bottom of the reservoir when the permafrost thaws. It may be necessary to consider placing an impervious film, such as clay or a rubber membrane around the bottom and sides of the reservoir to reduce seepage. The ice content of the underlying ground must be considered because thawing of permafrost in such materials will cause settlement and cracks in the dykes. Many communities are located on the sea coast and desalinization plants would be required if seawater were the only available source of water.

b / Distribution methods
The most common methods of water distribution in many Canadian communities in permafrost areas are by sledge and barrel or by tank truck. All of these subject the water to handling and possible contamination and are very costly. In summer it may be possible to distribute water under pressure through pipes laid on the ground. During the winter, however, such a system must be dismantled and drained, and the pipes left to lie on the ground.

Water pipes cannot be laid in the ground below the seasonally frozen zone as is done in southern areas free of permafrost. Where water distribution is required throughout the year, the mains have to be placed in heated utilidors (which have already been described above; see p. 69) or a recirculating distribution system must be installed to prevent freezing. Above freezing temperatures are maintained in utilidors by heat losses from the steam and condensate lines in the utilidor, by warm air forced through the utilidor, or by heating the water. Continuous circulation of the water is relatively easy to maintain but utilidors may cost from fifty to a hundred dollars per foot and their operation is expensive.

Utilidors may be built to carry several services such as steam, water, sewer communications, and power (Figure 26). Where sewer and water pipes are carried together, leakage of the sewer pipes may contaminate the water supply. This danger is reduced if the water pipes are carried above the sewer pipes. Drainage is a major consideration because flooding by groundwater may contaminate the water supply and ruin the insulation. Thus underground utilidors must be watertight and those above ground must be built so that they can be drained.

A recirculating water system consists of a distributing main, a return main, circulation pumps, and a water heating system. It may be a single main or dual main system: in the former, one pipe serves as both distribution and return main and is looped in one continuous line starting and ending at the recirculation pumps; in the latter, a two-pipe system is used with the high and low pressure lines side by side. If possible, waste engine heat from the circulating pumps should be used for preheating the water to several degrees above freezing.

In deciding, with either utilidors or a recirculation system, whether to place the pipes above or below ground, probably the most important factor is the condition of the soil and the difficulties that may arise from frost action in the active layer or from thawing of the permafrost. It has been suggested that a surface system should be used unless the active layer and underlying permafrost are in well drained, non-frost-susceptible soils extending to a depth of approximately fifteen feet.

If the services can be placed below the ground surface, the question of construction costs and heat loss become important. When placed above the ground, however, it is subjected to very low air temperatures in winter but also summer temperatures are high. If the system is in the ground, it may be in the permafrost or the active layer: below freezing temperatures prevail in the permafrost throughout the year but they are at no time as extreme as when the system is near the ground surface; in the active layer, advantage may be taken of the latent heat of fusion of entrapped water but in addition there is a danger of flooding from the

FIGURE 26 Utilidor at Inuvik bringing services to houses. Note pile foundation for house in foreground.

suprapermafrost ground water. When utilidors are constructed on frost-susceptible soils they are usually supported on piles with a minimum air space of two feet between the utilidor and the ground or placed on gravel mats several feet thick.

2 *Sewage Disposal*

In temperate climates, large amounts of waste material are reduced and destroyed through biological and chemical action of the soil. Excrement is placed in appropriate zones of the soil and is decomposed and rendered harmless by the complicated reductive forces of nature. The effectiveness and safety of most methods of sewage disposal in permafrost regions are greatly impaired by the low air and ground temperatures since permafrost, the extended period of seasonal frost, and long periods of low air temperatures retard biological and chemical reduction of organic material. These reductions do not appear to occur significantly in the perennially frozen ground and only very slowly in the top layer of the seasonally frozen ground where leaching of the soil is retarded and the poor drainage results in cesspools if sewage is left on the ground. This is liable to contaminate surface sources of water.

a / Primitive disposal methods
The use of pit privies and septic tanks as a means of sewage disposal is unsatisfactory, and many northern communities have to rely on the primitive pail system for disposal of human wastes. These wastes are simply dumped away from the settlement, possibly into a lake or river in summer or on the river ice in winter, to be carried away during breakup; garbage cannot be buried owing to the permafrost and hence in many cases is strewn on the ground around small villages; and waste water is usually allowed to drain into the soil around each building but discharging warm liquid wastes near a building will thaw the underlying ground causing the building to settle.

b / Collection systems
There are very few water-carried community sewage collection systems in permafrost areas partly because of lack of water and the problem and cost of combating low temperatures, frost action, and permafrost. Sewers, like water lines, can be kept operative only if they are protected from freezing, and differential movement owing to frost action in the seasonally thawed layer or thawing of the permafrost.

Enclosing the sewer in a heated conduit or utilidor is the best way of ensuring trouble free operation but these installations are expensive. If the pipes can be placed in the ground, the high cost of utilidor construction may be avoided. The feasibility of such buried installations is governed by soils and permafrost conditions. With both buried and above-ground lines, adequate supports to maintain proper alignment of pipes are essential and in fine-grained soils with high moisture contents subject to thaw settlement, pilings anchored in the permafrost may be required.

Because sewage contains warm waste water from heated buildings its temperature is generally higher than the water supply and less likely to freeze. Nevertheless, precautions must be taken to guard against freezing. Pipes below the ground should be located to take advantage of solar radiation and the insulation effects of snow and vegetative cover.

In most northern communities, the final disposal of sewage is simply by dilution in large water courses or tidal waters. For isolated communities this method is satisfactory but if other water users are relatively close downstream or along a coast with coastal currents, serious health hazards can be created. Even with the present sparse population, stream pollution has already posed health problems.

c / Sewage treatment
Sewage treatment is attempted in some communities but there are difficult problems. In the southern fringe of the permafrost region there is concern where small private water supplies and sewage disposal installations, like cesspools for single dwellings, are located near each other. There is little information on the movement of polluted material underground in such a situation during the summer when the ground is thawed for some depth. The fact that the conventional sewage treatment processes require a warm environment to function properly complicates the design and construction of plant facilities in permafrost areas and, combined with the expense of providing the necessary heat, adds greatly to their cost. It is possible to provide septic tanks in heated buildings at the outflow of the sewers but heating expenses add to the cost.

Increasing attention is being paid to the method of sewage treatment called lagooning whereby raw sewage is dumped into large shallow lagoons or oxidation ponds, as they are sometimes called, where aerobic conditions are maintained, and natural decomposition takes place. Sewage lagoons provide for biological or secondary treatment of sewage and, if properly constructed, provide treatment approaching that in most of the more expensive treatment plants. They appear to work satisfactorily in cold climates, although their effectiveness is reduced in winter. Depth should not exceed three to five feet to ensure aerobic conditions and eliminate odours, but in cold climates greater depths are necessary to allow for four feet or more of ice cover.

The initial and operating costs of lagoons are less than for conventional treatment plans. An important consideration with lagoons is the retention of sewage in areas of soils with high ice content where thawing of the permafrost can cause settlement. Here dykes may wash out and so allow seepage of sewage wastes into

the surrounding area. In many respects, sewage lagoons appear to offer an economical and acceptable answer to the problem of sewage treatment in permafrost
areas.

3 *Experience in Northern Canada*

Settlements in Northern Canada obtain water from various sources, including
lakes, streams, even one from the ocean, and wells. Pipe systems exist in the large
communities but the more primitive distribution methods are employed in the
smaller settlements. Modern sewage systems are also in use in large centres but
the smaller places rely on primitive disposal methods. Various techniques have
been used in the design and operation of water supply and sewage disposal
systems to counteract adverse soil conditions and permafrost. Considerable
success has been achieved in settlements in both the discontinuous and continuous
permafrost zones and some examples are described here (Yates and Stanley
1966).

a / Discontinuous permafrost zone
A number of wells have been drilled successfully in this zone. At the motels at
Mile Posts 1,033 and 1,095 on the Alaska Highway, 100 miles from the Alaska
boundary, artesian wells have been developed by drilling for 160 feet, approximately 90 feet of which is through permafrost. These wells can be kept flowing
either with an electric heating cable or by permitting continuous flow. At Rae,
north of Great Slave Lake, a well was drilled in unfrozen ground to a depth of
32 feet.

At Thompson, water and sewer mains were buried at least 10 feet deep for
protection from seasonal frost penetration (Klassen 1965). In permafrost areas,
trenches were excavated several feet below the pipe grade and backfilled with
sand to the ground surface to minimize settlement of the mains. Heating of the
water in the distribution system is required from January to the middle of May.
Recirculation is normally started several weeks before the heat is turned on and
continued for some weeks after it is turned off until all danger of freezing has
passed. In areas where permafrost was not completely removed, subsidence causes
the pipes to settle and sometimes break. The usual practice is to allow the pipes to
settle until they break or until subsidence is complete, after which the pipelines
are repaired and regraded. Water mains may be raised above grade appreciably, in
these areas, to allow for some additional settling. At the boundaries between
permafrost and non-permafrost areas main breaks are a common problem, while
in permafrost areas mobile steaming facilities have been necessary to thaw frozen
mains.

Pipe distribution systems are used at Hay River and Fort Simpson where the distribution of permafrost is similar to that at Thompson, being in the form of scattered islands 25 to 50 feet thick. At Hay River water distribution and sewage disposal systems are buried in both the old and new townsites. At Fort Simpson water for the school, hostel, and hospital is obtained from two or three drilled wells and distributed in a utilidor system.

The town of Yellowknife has a population of approximately 4,000 people about 2,700 of whom are served by a waterworks and sewerage system constructed in 1947 and 1948 (Stanley 1965) on the townsite, a flat sandy area where the existence of scattered patches of permafrost added greatly to the cost of construction. In the surrounding area, permafrost reportedly exceeds 150 feet in thickness.

Water from Great Slave Lake is piped through mains laid at a minimum depth of six feet below the ground surface and insulated with moss. This recirculating, dual-main distribution system consists of two separate systems of mains and service connections: water flows from one system to the other through an orifice connection in the basement of each building and returns to the water plant through the second system. The entire distribution system is graded so that it can be drained in the event of a prolonged breakdown. From early November until late May the water is heated to 40 °F to prevent the water freezing because of low ground temperatures (Grainge 1959). The sewerage system is buried close to the water lines to prevent freezing. Sewage is collected and pumped into a small lake, which is used as a sewage lagoon or storage pond.

Norman Wells obtains water from Bosworth Creek which flows into the Mackenzie River immediately downstream from the settlement. Underground utilidors carrying steam, condensate return, water and sewer lines at shallow depths (four to five feet) were used first during the Second World War, but unsuccessfully, owing to ground settlement and freezing. They were replaced by wood frame utilidors above ground which were weatherproofed, insulated with asbestos paper, and set on short posts spaced approximately 8 feet apart and driven into the active layer (Grainge 1959). Shortly after installation, however, some of the posts were heaved upward as much as 20 inches by frost action. The utilidors were then laid on steel piles which were reportedly driven 10 to 15 feet into the permafrost at a spacing of 7 to 9 feet (Hemstock 1953), but because some of the piles were not anchored securely into the permafrost, some heaving and settlement still occurred. During the summer of 1954 the utilidors were laid on small wooden blocks on a gravel pad; these pads proved easier to maintain after differential settlement than the piles (Grainge 1959).

Dawson lies at the junction of the Yukon and Klondike Rivers in central Yukon Territory. Its population (about 500) uses water pumped from the Klondike River and supplemented by wells. In 1904 there were put into operation waterworks and sewerage systems constructed of wood stave pipe laid in gravel in the seasonally thawed layer above the permafrost table. The water system was laid approximately 4 feet deep, although in some places it is as shallow as 6 inches deep, and the sewer system in general is not shallower than 4 feet (Stanley 1965). In the winter, frozen ground surrounds the pipes, but frost heaving damage is prevented by good drainage in the gravel and the water is heated by electricity to 42° F. Sufficient water flow to prevent freezing is maintained by bleeding in each house and at the dead ends of the mains. At the end of the circulation system the water temperature is 34° F (Grainge 1959).

At Clinton Mine, 50 miles northwest of Dawson, the permafrost is also widespread and several hundred feet thick. All services at the plantsite and townsite except sewerage are carried in a utilidor – 47-inch diameter steel pipe available at salvage prices from the Klondike area (Drewe 1969). The pipe was laid in a trench excavated in the active layer and backfilled with compacted gravel so that the top of the pipe was just below the final ground surface. It was felt that this pipe would provide rigid casing for the steam, condensate, water and fire lines which could not be affected by possible thaw settlement in areas of silt and high ground ice content. Sewer lines are of asbestos cement pipe laid in a trench in compacted gravel at a depth of 6 feet. In the townsite where the steam and water lines branch off the main utilidor to service several houses, the steam, steam return and water pipe were laid side by side in a trench excavated 2 feet into the active layer and backfilled with compacted gravel. One break has occurred in the sewer line which was attributed to the presence of permafrost.

The town of Schefferville, P.Q., lies in the discontinuous permafrost zone in central Labrador-Ungava but no permafrost has been found in the townsite. In providing water and sewage disposal facilities the main problem is the cold winters, with deep frost penetration. This same situation prevails at Uranium City on Lake Athabasca, in northern Saskatchewan where all water and sewer pipes are buried to a minimum depth of 9 feet except in several locations where since it was impossible to trench to 9 feet the water mains were insulated. Steam thawing equipment and electric transformers have been required periodically to thaw pipes (Klassen 1965). At the southern boundary of Mackenzie District on the Slave River is Fort Smith. In some places in the area there are indications of islands of permafrost but generally when areas have been developed, the permafrost has receded (Stanley 1965). Water is drawn from the Slave River, heated and treated, and distributed through a single main system. In 1950 the pipes were laid at a depth of 9 feet below ground surface and the householders at the ends of the lines kept taps running slowly, day and night to prevent the water from freezing. These pipes have since been relaid to a depth of approximately 12 feet, that is, well below the maximum penetration of seasonal frost (Grainge 1959). The sewerage system is of conventional design and is buried at 9 feet. As in the example of Uranium City the deep burial of water pipes results in higher costs for installation and maintenance. Another problem was how to build a supply line from the intake structure up an unstable hillside to the water treatment plant when there appeared to be perennially frozen ground in the hillside, and each summer there was slipping along the surface. A buried pipe or piles was impractical, so a pipeline was designed using 6-inch cast iron flexible joint pipe with styrofoam insulation and several expansion joints at different locations on the hill. The pipe between expansion joints is held together by cables and the supports at each joint hold the pipe above the surface to allow longitudinal movement. After movement takes place in the hill, re-adjustments can readily be made to the pipeline (Stanley 1965).

Whitehorse is situated on the Alaska Highway in southern Yukon Territory. Prior to 1956, a small portion of its population was served by the army water supply system whereby unheated water from a nearby creek (which does not freeze to the bottom in winter) was pumped to a main feeder line at a depth of 6 to 7 feet below the ground surface. In 1956 a new system was installed to pump water from the Yukon River and wells through a distribution system buried at a depth of 9 feet below the ground surface where, unlike the army system, it was almost completely safe from seasonal frost during severe winters. Although no permafrost has been encountered in the townsite, islands do exist in the surrounding area. Because of the severe winters and great degree of frost penetration, more excavation and backfilling is required at Whitehorse than at settlements to the south causing higher costs in placing and maintaining a water distribution system (Copp *et al.* 1956).

The construction and operation of these few water and sewage systems in the discontinuous permafrost zone are conditioned by local factors such as the distribution of permafrost, its depth below the ground surface, and temperatures in the annually thawed and frozen layer above the permafrost. At Schefferville, Uranium City, Whitehorse, and Fort Smith, where no permafrost has been encountered in the townsite areas, the mains have been placed below the depth of seasonal frost penetration. At Thompson some permafrost is excavated during installation of pipelines, at Yellowknife permafrost is scattered in the townsite and lies several feet beneath the maximum seasonal frost penetration, allowing placing of the mains between the two layers of frozen ground whereas Dawson and Norman Wells are situated further north where permafrost near the ground surface requires specially modified distribution systems.

b / Continuous permafrost zone
Inuvik, on the east side of the Mackenzie Delta, is situated on soils ranging from fluvioglacial gravel to till. Water is supplied to its population of about 1,500 from a nearby lake into which river water from the East Channel is pumped when the water in the river is clear. Water and sewage mains are carried in in-

sulated utilidors supported above ground on piles extending to a minimum depth of 10 feet (Figure 26). Sewage is carried (see Figure 27) by gravity to a lagoon located one mile west of the town where it is discharged and subjected to treatment before draining into the East Channel of the delta (Pritchard 1962).

Aklavik obtains water from early June until late September from a nearby lake, through a 2-inch diameter pipeline laid on the ground surface; it is pumped through a diatomaceous earth filter and then chlorinated before distribution. During the winter, ice blocks are cut from the Peel Channel and distributed by sledge. An unprotected pipeline cannot be laid either above or below ground in winter because of freezing temperatures and the expense of constructing an utilidor system cannot be justified by the the total consumption. Raw sewage is collected from individual buildings and dumped into the Peel Channel or on the river ice in winter to be carried away during breakup.

In the small settlement of Tuktoyaktuk (situated on the arctic coast eighteen miles east of the East Channel of the Mackenzie Delta) water is hauled from a nearby lake in summer. Water from the Beaufort Sea can be used during the winter because of a favourable natural phenomenon. After the sea begins to freeze, in late October, the fresh water of the Mackenzie River spreads out over the heavier salt water, displacing it downward with very little mixing of the two. At the

FIGURE 27 Twisted sewage disposal line at Inuvik caused by downslope movement of thawing, perennially frozen, granular (coarse-grained) soils which were exposed in sidehill cut.

end of October only one to two inches of fresh water have accumulated below the ice surface but by late December this layer is sufficiently thick for fresh water to be piped from the sea and by breakup time in late May the water is fresh for a considerable depth. Sewage from government housing is dumped in a natural coastal lagoon, which is cleaned out periodically during heavy storms.

A line was constructed in 1961 at Churchill, Manitoba to supply water from the Churchill River to the town of Churchill and the military establishment, Fort Churchill (Engineering News Record 1962). This new line replaces pipes laid on the ground surface which brought water from a shallow lake. It consists of a 6,800-foot long 16-inch diameter steel pipe intake siphon from the river, a 28,000-foot long 10-inch diameter aluminum pipeline, which runs to a pumping station between the two centres, and a 7,000-foot long line of 6-inch diameter aluminum pipe connecting two existing distribution pumping stations. The 10-inch line was laid in a trench 4 to 6 feet deep of which all but 1,700 feet was excavated in permafrost. Excavation began in June, 1961 and the frozen ground was removed by stripping the surface vegetation and the underlying soil in layers as it thawed. In some cases six passes were required with a dragline to dig the trench to the 6-foot depth. Excavation of the trench for the 6-inch line was accomplished by blasting the perennially frozen ground. The pipe was insulated with a ½-inch thick layer of cork mastic to prevent freezing. No problems have been reported with permafrost in the operation of this line (Personal communication, Department of Public Works).

A water treatment system was constructed at Frobisher Bay in 1963 at a lake on a hill above the town (Personal communication, Northern Canada Power Commission). A utilidor leads down to the hospital and from there a sanitary sewer extends around the new housing development and directly to the bay. There is no water-piping system in the settlement. Many houses have water and sewage tanks which are serviced by tank trucks. The Department of Indian Affairs and Northern Development has installed a small experimental pump circulating sewage system, with holding tanks outside each house, which serves about 15 houses near the bay. This system has been performing satisfactorily.

A 6-inch steam line with 3-inch condensate return line above ground was built on timber frames and rock filled timber cribs to serve the large Federal Building. This line is equipped with expansion loops about 10 feet long and 5 feet wide to allow for ground movements due to permafrost thawing and frost action. This loop arrangement is less expensive than expansion joints normally used, for example, at Inuvik. The existing utility systems at Frobisher Bay are located mostly on rock or rock fill with a few poorly drained areas; no special precautions have been taken against permafrost and no difficulties encountered.

The settlement of Resolute in the Arctic Islands is typical of the many small settlements in the North which have no water and sewage systems. The buildings at the airport have a water supply system from which water is hauled by tank truck to a large storage tank in the settlement garage (Personal communcation, Department of Indian Affairs and Northern Development). The inhabitants obtain water from there in buckets as there is no distribution system. Some staff houses have water tanks. Several water point buildings (heated buildings which house water tanks) are being installed this year. Sewage is handled with plastic bags in chemical toilets. It is taken to a special area from which it is hauled onto the sea ice in winter and inland to a shallow depression in summer. No special problems have been caused by the permafrost in this far northern settlement where the active layer is only about one foot. The construction of a water distribution and sewage disposal system at Resolute would be very expensive and the provision of this service is improbable in the near future.

In the continuous permafrost zone few settlements have water and sewage systems because of the high cost of installing and operating them. Owing to the thickness of the permafrost, it is impractical to drill wells for water. It is not possible to find unfrozen ground in which services can be buried as at some settlements in the discontinuous zone. The best way to distribute water is through utilidors above ground but the cost is very high and as a result provision of these services can only be justified in a few of the largest centres.

B ELECTRICITY

1 Power Generation and Transmission

In the early days the only available source of electricity was from wind charger systems which provided DC power. This method was used widely in the North by church missions, the Hudson's Bay Company and others. Only a few such systems probably remain in service today and virtually all settlements obtain their electric power from local diesel engines or, more recently, turbine-driven generators. The Northern Canada Power Commission or Department of Indian Affairs and Northern Development have taken over the production of power in practically every settlement. Most of the settlements are sufficiently small that their requirements are satisfied by small units. A few of the larger settlements, however, such as Inuvik, Norman Wells, Hay River, and Frobisher Bay have greater power requirements, needing large equipment and hence larger structures to house this equipment. In many cases both heating and electrical generating equipment is housed in the same building thus imposing higher heat and structural loads. They require special consideration with regard to foundation design in permafrost areas.

The foundations for the generating equipment are in fact usually separate from the building foundation.

At large settlements where mining operations are underway, the large quantities of power required are obtained from hydro-electric plants (Polar Record 1967). Special problems associated with permafrost arise in the design, construction, and operation of these structures and one of the most critical of these problems results from the impounding of large bodies of water over perennially frozen ground. The thermal effect of the water causes thawing of the underlying permafrost which may adversely affect the performance of water-retaining structures such as dams and dykes. Another problem is encountered in obtaining and handling perennially frozen earth fill material. During the past few years some experience in these problems has been obtained in northern Canada.

In addition to permafrost problems arising at the hydro-electric generating station, consideration has to be given also to the transmission of power to the consumer. The construction of transmission line towers may be quite difficult in areas where foundation soils are perennially frozen and contain large quantities of ice.

a / Experience in northern Canada
One of the earliest hydro-electric generating plants constructed in the permafrost region, and the first after the Second World War, was the Snare River Power Project operated by the Northern Canada Power Commission. This plant is located ninety miles northwest of Yellowknife in the discontinuous zone where permafrost is widespread; it was built to supply power to Yellowknife and the rapidly developing gold mines in the area. Construction began in 1946 and the first power was transmitted two years later. The main dam consists of a rolled impervious rock flour core, and pervious embankments grading from fine sand adjacent to the core, to quarried rock at the outer faces (Eckenfelder 1950). The core material, obtained from nearby swamps, was perennially frozen with considerable quantities of ice. This necessitated costly and time-consuming treatment before it could be placed. The procedure followed was allowing the pit to thaw naturally to a depth of six inches or more, after which the unfrozen material was scraped off by bulldozers and pushed within reach of a dragline standing on firm ground at the side of the pit. The bulldozers, as long as they did not travel in the same track repeatedly, had no difficulty moving on the freshly uncovered, frozen, fine-grained soil and ice lenses. If they travelled too much in one area they became hopelessly bogged down with resulting frequent transmission failures. Because the dam was built on permafrost it was certain to settle considerably once the thawing effect of the reservoir water was felt. Rather than attempt to correct this condition, the dam was simply built about ten feet higher than required thus permitting a large amount

of settlement. Wide rock berms were provided to prevent pushing out of the under-lying material due to excessive settlement. The fill actually settled about three feet during the construction period, but subsequent movement was negligible (Eckenfelder 1950). A second plant was built in 1960 a few miles away to supply Yellowknife's growing needs.

The Kelsey Generating Station of Manitoba Hydro is located in northern Manitoba on the Nelson River about 400 miles north of Winnipeg. It was con-structed to supply power to the International Nickel Company of Canada Limited at Thompson, Manitoba, approximately 50 miles to the southwest. Construction began at Kelsey in 1957 and was completed in 1960 (Hopper 1961; MacDonald *et al.* 1960; MacDonald 1966) at the northern end of glacial Lake Agassiz where the predominant soils are lacustrine varved clays which may exceed 25 feet in thickness. These deposits overlie bedrock or thinly stratified glacial drift. The drift, from zero to more than 20 feet thick, is composed predominantly of a sandy gravel or medium to coarse sand. The total thickness of overburden varies from a few feet to more than 50 feet. The mean annual air temperature is about 25° F.

Kelsey is located in the middle of the discontinuous permafrost zone. Permafrost is widespread, occurring as scattered islands with a maximum thickness of possibly 50 to 60 feet. The mean ground temperature in the permafrost varies from about 30.5° F to 31.5° F. The frozen clay contains large quantities of ice, generally in the form of horizontal lenses from hairline to 8 inches thick but averaging 1/16-inch thick.

In order to cope with these permafrost conditions, special designs had to be formulated for two of the major dykes, particularly to accommodate the large amount of ground settlement as the perennially frozen ground thawed. Briefly, the design consisted of a compacted sand fill constructed on the frozen varved clay that is stabilized during thawing by a drainage system of sand piles. As the dykes settle, new fill is added to bring the crests to the required elevation (Figure 28). The dykes have performed as anticipated and there has been no interruption in service. Some difficulties were encountered with permafrost during construction of the sluiceway and clay dykes between the sluiceway and the powerhouse. Initial attempts at excavating the spillway channel consisted of stripping the upper sur-face down to the permafrost table and allowing the sun to thaw the frozen soil. This resulted in equipment becoming bogged down in the resulting quagmire. Various types of drilling and blasting were then tried but the resultant lumps were too large to handle. The use of large power shovels finally proved to be the most successful method of excavation and it was by this means that the earth excava-tion was finally completed (Hopper 1961).

As mentioned above, the thawing effect of water on permafrost is of particular concern in the design and construction of water-retaining structures placed on perennially frozen ground but there was little practical experience or information available on the problem. A co-operative study was initiated in 1958 by the Division of Building Research, National Research Council of Canada and Manitoba Hydro to assess the performance of two major, but relatively small (2,000 feet long, maximum height 20 feet), dykes constructed on perennially frozen ground. Preliminary analyses, though only approximate, indicated that within a 50-year period significant thawing would occur beneath the dyke-water interface and that the permafrost under the reservoir would completely thaw. Ground settlements of as much as 6 feet under the dykes were to be expected as thawing took place. To check the estimated performance, field instrumentation was installed and an observational programme begun in 1959 (G. H. Johnston 1965, 1969). Flooding of the reservoir was begun early in 1960 so that installations in the forebay were covered during that summer. Water impinged on the dyke at two of three instrumented locations late in 1960. In the fall of 1962, the water level was raised to its maximum elevation and thus has been in contact with the dyke at the third instrumented location from that time. By 1966 observations indicated that the permafrost had thawed to a depth of about 15 feet under the forebay. Initially the rate of thaw was about 5 feet per year but this has decreased to approximately 2 feet per year. Similar depths and rates of thaw have been observed in the dyke foundation. Permafrost has thawed completely at one location on the dyke. The maximum settlement of the ground surface observed at the instrumented locations is almost 4 feet. Settlements of as much as 7 feet have occurred locally, however, at other areas on the dykes. A distinct pattern of thawing and settlement each year has been observed. Major thawing and settlement occur during the summer

FIGURE 28 Dyke at Kelsey hydro-electric generating station in discontinuous permafrost zone in northern Manitoba. Lower dark portion of crushed rock, which is original dyke, settled because of thaw settlement of underlying, ice-laden, perennially frozen varved clays. Top lighter lay of gravel was added to bring dyke level up to original height.

months at rates that are much greater than those experienced during the winter. The change in rate of thaw (and thus also the rate of settlement, which is partly a function of thawing) is directly connected with the change in water temperature or rate of heat flow into the ground. During the period November to May, the mean water temperature is very close to 32° F but during the summer months it increases rapidly with maximum water temperature of from 60 to 70° F occurring during July and August. The mean yearly water temperature is about 43° F.

The Kelsey Generating Station was the first dam built on permafrost in northern Manitoba. In the early 1960s an intensive study was initiated of the hydro power potential along the Nelson River from Lake Winnipeg to Hudson Bay. In 1966 the Federal Government and Manitoba Hydro announced the establishment of a 30-year programme of power development on the Nelson River. Construction of the first hydro-electric plant after Kelsey began in 1966 at Gillam, about 150 miles south of Churchill. It is located in the northern part of the discontinuous permafrost zone where permafrost reaches depths of about 80 to 100 feet. In the construction of this project and others to follow, permafrost problems similar to those at Kelsey will probably be encountered.

Construction of a hydro-electric plant began recently in southern Labrador at Churchill Falls on the Churchill River. Being several hundred feet high, the falls is one of the largest potential power sources in North America, possibly in the world. It is located in the southern fringe of the permafrost region where permafrost occurs in scattered islands. Here a few scattered islands of permafrost have been encountered during the early stages of construction (Personal communication with R. N. Seemel, Acres Canadian Bechtel). During the construction of another plant in Labrador at Twin Falls, near Wabush, about 1960, permafrost was encountered at the bottom of the canyon, 300 feet deep. It is likely that its existence was due to the shading in the depths of the canyon (Andrews 1961). No further reports of permafrost have been obtained from this site.

In addition to posing a problem at the dam sites, permafrost is an important consideration in the construction, design, and maintenance of tower foundations on the transmission lines. Many towers will have to be constructed on perennially frozen ground, much of it varved clays with high ice content. The Division of Building Research, National Research Council of Canada, in co-operation with Manitoba Hydro, initiated a programme in 1966 to study the problem of providing adequate anchorage for EHV guyed transmission line towers. Test sites have been established at Gillam and Thompson where various types of anchors have been installed in perennially frozen soils.

Even small power distribution lines in settlements require special consideration because of the permafrost. A line at Inuvik, extending eight miles to the airport from town, was erected in a different manner. At each pole location, a pile, called a stub, was placed in a hole excavated 12 feet into the permafrost and extended level up to original height.

3 feet above the ground surface. A pole was then strapped to the pile (see Figure 29). This line has performed satisfactorily since its construction.

As new settlements develop in northern Canada, the production of hydro-electric power will be an important consideration. Permafrost will impose problems at most sites requiring special modifications in design and construction of dams, dykes, and other structures to suit the particular conditions.

2 Communication Lines

Telecommunication facilities in northern Canada are limited but they are increasing steadily with the growing demand for improved services. Several telephone and telegraph lines extend into the permafrost region now – along the Hudson Bay Railway to Churchill, along the Mackenzie River system to Yellowknife and Inuvik, and along the Alaska Highway. Radio towers and masts have been constructed at most settlements and Loran towers for long range air navigation exist at far northern points such as Cambridge Bay, NWT in the Arctic Archipelago.

Various techniques have been employed to cope with adverse soil conditions and permafrost. The line to Churchill consists of tripod poles placed on the ground surface to avoid the expense of excavating the frozen ground and to minimize

FIGURE 29 Line pole strapped to stub imbedded in permafrost at Inuvik.

heaving to which they are more tolerant. Ground surface movements due to frost heaving have moved some of them but no line damage has been reported. The same design was used for the line to Yellowknife.

About the end of the Second World War a Loran tower 600 feet high was erected at Kittigazuit, 60 miles north of Inuvik in the continuous permafrost zone. It operated for about three years and then it was removed from service because of transmission difficulties. In 1955 the tower was dynamited and toppled to the ground because it was considered a hazard to aircraft. It was constructed on a sandy fill, several hundred feet in dimension, rising about 15 to 20 feet above the tundra surface. The tower rested on 4 concrete footings, 20 feet square, on steel piles embedded in the permafrost. The top surface of the footings extended about 2 feet above the fill surface. Guy wires were anchored to concrete blocks about 500 feet from the tower.

Special provisions were made in the design to spread solar heat transmission from the tower over the ground surface and thus minimize heat flow into the ground which would thaw the permafrost under the footings. A network of copper wires, both above and below the ground surface, extended out from under the fill – presumably from the steel piles under the footings – to copper rods several hundred feet away. This foundation design appeared to be satisfactory although the tower was in operation for only a short period of time. Levels taken on the footings five years after it ceased functioning showed no movements and there was generally no serious deterioration of the concrete.

The Department of Transport is responsible for the construction and operation of many telecommunications structures in Canada including the permafrost region. It is generally the policy of the Department to design and construct these foundation structures to carry the design loads with a minimum of maintenance and considerable success has been achieved in satisfactory performance (Sebastyan 1963). If soils are well-drained, coarse-grained, and not frost susceptible, spread foundation structures of wood, steel, or concrete can be placed in the active layer. If the soils are fine-grained and frost-susceptible, wood or steel piles are frequently used, installed in the permafrost to at least twice the expected maximum depth of the active layer during the lifetime of the structures. The type of piles, numbers and spacing are determined on the basis of loads imposed by the telecommunications structures and anchoring guys.

C   UNDERGROUND STORAGE

The below freezing ground temperatures in Canada's permafrost region have been used to a limited extent in preserving perishable foodstuffs. The usual present day

means are by mechanical refrigeration plants but the natural cold of the permafrost can provide suitable underground storage if adequate measures are employed.
There are in fact some advantages in using natural storage rather than artificial
freezing. The chambers are easily kept at uniform temperature and moisture
conditions. Artificial refrigeration plants require skilled operators. There is little
or no danger of fire in permafrost storage areas. Construction and maintenance
costs are low. In the past only the storage of food has been considered but in the
future the permafrost may be utilized for the storage of petroleum products and
natural gas.

The earliest reference to the use of natural cold storage in northern Canada
appeared in 1866, when large quantities of meat and fish were being kept in
excavations in permafrost along the lower Yukon River. By 1885 this practice
was in use at Point Barrow, Alaska. Shortly thereafter, American whalers established cold storage caches along the Canadian arctic coast at various points, such
as Herschel Island, Cape Bathurst, Cape Parry, and on Banks Island. When the
whaling fleets withdrew, many of these cold storage chambers fell into disuse, but
this method of storage was adopted where soil conditions are favourable for
excavation (Copland 1956).

1 Construction and Maintenance

The two types of cellars used were vertical excavations and horizontal tunnels.
Vertical shafts were excavated in winter when groundwater in the active layer
is frozen and would not flood the hole. The shaft was usually about 20 feet deep
and widened increasingly below a depth of 6 feet, so that a storage space of perhaps 8 to 14 feet square was formed at the bottom. Logs were used to frame the
shaft and double trap doors built into the entrance. The wider part of the shaft
was divided into 2 storeys, the upper part being used to freeze cuts or carcasses
individually before they were placed in the lower chamber for storage. A slanting
roof could be built over the site and extended to form an entrance porch in which
double doors were built. The roof was covered with an insulating layer of turf
approximately 6 inches thick, and the excavated material heaped over the structure. A horizontal tunnel was excavated more easily than a vertical shaft. It was
difficult, however, to prevent the escape of cold air from the tunnel in summer,
and therefore temperatures were not sufficiently low for storage beyond one or
two summers (Copland 1956).

Several factors limit the wide use of natural cold storage: excavation had to be
in winter when working conditions were difficult; and during the summer the air
temperature in the cellar should be maintained below 15° F, to prevent the growth
of mould on stored meat. The upper part of the cellar might be in the seasonally

thawed or active layer, and as thawing progressed through the summer, seepage could occur in the excavation; because there is a lag between surface air temperatures and soil temperatures, thawing could continue for several months. To overcome this, it was advisable to cover the excavated area with moss or peat to raise the permafrost table above the top of the cellar. Drainage of the surrounding area was also important and care was required to ensure that the entrance to cold storage rooms was higher than the surrounding ground to prevent seepage of water into the underground chamber (Copland 1956).

2 *Experience in Northern Canada*

Natural cold storage was used in the Aklavik area where the fine-grained, uniform, deltaic soils made excavations relatively easy. The Anglican and Roman Catholic missions had deep cellars in permafrost to store the several tons of reindeer meat for their hospitals and boarding schools. In addition to the storage of meat and fish in Aklavik, river ice cut in winter was kept underground for summer drinking water. At Reindeer Depot it had been impossible to keep fresh meat from the annual slaughter before 1940 when a cold storage chamber, large enough to hold 100 reindeer carcasses, was excavated in perennially frozen silt. At Old Crow on the Porcupine River in northern Yukon Territory there were cold storage cellars in the perennially frozen alluvial silts and sufficiently low temperatures were maintained to ensure the preservation of meat (caribou and moose), vegetables, and canned goods indefinitely (Leechman 1948).

A number of settlements on the western arctic coast used cellars excavated in permafrost: the Hudson's Bay Company built one to store ice in 1928 at Cape Bathurst; the cold storage cellars excavated by American whalers at Herschel Island in the nineteenth century were used by the Hudson's Bay Company and Royal Canadian Mounted Police for storing ice; and a "freezing house" was constructed on Nicholson Peninsula by removing three feet of overburden and chiselling out the frozen ground beneath to a depth of six feet (Copland 1956).

On the other hand, east of Cape Parry there was little storage in permafrost because bedrock and glacial drift with boulders made excavation difficult and expensive. There were cold storage cellars at Eskimo Point on the west coast of Hudson Bay, one of which, built by the Anglican Mission, was covered with peat and freezing temperatures were maintained throughout the year. On the other hand, an underground ice house built by the Royal Canadian Mounted Police was damaged by the penetration of heat and water because the walls were not insulated (Copland 1956).

There are now virtually no underground cellars in permafrost in Northern Canada (Personal communication, Hudson's Bay Company). Practically all settlements in the Arctic have electricity and are equipped with refrigeration plants. A small cellar is still used at Repulse Bay in the central Arctic where there is no electricity. It is only five feet square, extending six feet below the ground surface into the permafrost and covered by an enclosure above ground. A larger community cellar has been in use at Coppermine but its operation may have ceased. At Tuktoyaktuk a large cellar, located under a building, was used for many years but it is probably not functioning now. The use of permafrost for natural cold storage of perishable foodstuffs in northern Canada has always been limited and the accounts presented here are mainly of historical interest. Nevertheless, new aspects of the utilization of permafrost may arise in the near future with consideration of underground storage of petroleum and natural gas in the Arctic.

D CONCLUSION

Because the population of the permafrost region of Canada is sparse, there are only a few settlements which are sufficiently large to require or be capable of financing elaborate systems for providing services. The structures which have been constructed to distribute these services, such as utilidors, are very expensive and special design features are required to meet the problems caused by permafrost. In the many small settlements, water supply and sewage disposal are handled by simple methods and electricity is provided by small generating units. Even here some water distribution and sewage disposal systems are planned and under construction. The natural cold storage possibilities provided by the permafrost are a special feature of these cold regions. Although limited use has been made of this feature in the past, there may be considerable use for underground storage in the future.

6 Transportation

From the first years of activity in Canada's permafrost region, transportation has provided the key to exploration and economic development. In the early days of the fur trade around Hudson Bay and in the northwest, travel was along water routes and overland on foot. Permafrost presented no problems, and in fact its existence was hardly known or suspected. The first land route into northern Canada where the existence of permafrost was mentioned and documented as a problem was the Hudson Bay Railway, completed in 1929 from The Pas to Churchill to carry Prairie grain to Hudson Bay. The White Pass and Yukon Railway connecting Whitehorse, YT with tidewater at Skagway, Alaska was built several decades earlier but problems with permafrost have not been evident in the construction or operation of the line.

The Mackenzie River system has long served as a transportation route from the Prairies to the arctic coast. From Waterways, 300 miles north of Edmonton, the system provides a 1,600-mile navigable waterway to the Arctic Ocean, with only one interruption – 12 miles of rapids on Slave River between Fort Fitzgerald, Alberta, and Fort Smith, NWT, at the sixtieth parallel. Today it is still of vital importance, carrying the bulk of freight into Canada's Northwest.

In the early 1940s transportation facilities were expanded in Canada's Northwest in response to the defence emergency created by the Second World War. Three large projects were completed in great haste with little regard for cost: the Alaska Highway extending 1,500 miles from Dawson Creek, BC to Fairbanks, Alaska; the Northwest Staging Route comprising a network of airstrips in Northwestern Canada and Alaska; and the Canol Project involving the construction and operation of a 620-mile oil pipeline from Norman Wells on the Mackenzie River to the Alaska Highway in the Yukon Territory. Because the engineers and contractors involved in these undertakings in Yukon Territory and Western Mackenzie District were unaware of the existence of permafrost, construction methods similar to those used in temperate regions were employed. As a result, permafrost caused many unanticipated problems. These three projects were instrumental in awakening Canadian and American interest both in the North and in permafrost.

Since 1945 transportation facilities in the permafrost region have been expanded considerably in response to economic developments (as seen in Figure 30). The Canol Road, except for a portion in Yukon Territory, has been closed but the Alaska Highway has been expanded into a network of roads called the Northwest Highway System. The Mackenzie Highway was built in the late 1940s through Northern Alberta to Hay River, and connecting roads have been constructed to Yellowknife. New roads are presently under construction to Fort Smith and Fort Simpson. In recent years the highway systems of Saskatchewan and Manitoba have been extended into the southern fringe of the permafrost region.

Several railways have been built into the permafrost region to reach newly developed mining areas: the Quebec North Shore and Labrador Railway extends from the Gulf of St. Lawrence to the iron mines at Schefferville in central Labrador-Ungava, with a branch line to Wabush and Labrador City; in northern Manitoba two branch lines have been built from the Hudson Bay Railway – one extends to the base metal mine at Lynn Lake and the other to the nickel mining and smelting plant at Thompson; the Great Slave Lake Railway, completed to Hay River and Pine Point, now carries base metal ore concentrates to British Columbia; and the White Pass and Yukon Railway has provided a link for many years through the Alaskan Panhandle between Whitehorse and the Pacific Ocean.

Airfields have been constructed at several northern localities including weather

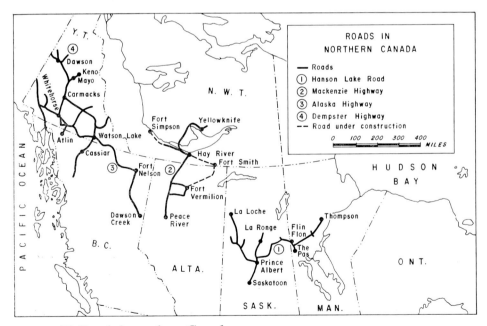

FIGURE 30 Roads in northern Canada.

stations on the Queen Elizabeth Islands and Distant Early Warning (DEW) Line stations. In many instances permafrost has created problems which have contributed to increased cost of construction and maintenance over that spent for similar projects in southern Canada.

Despite the construction in recent years of roads, railways, and airfields, transportation facilities in northern Canada are limited. Transportation is the key to development and there is strong argument that the construction of roads and railways into new areas will stimulate exploration and exploitation of suspected minerals and other resources. First, there must be economic reasons for such routes. At present, there are few markets in Canada's permafrost region and products from the north have to compete with those produced in temperate regions.

A   CONSTRUCTION AND MAINTENANCE TECHNIQUES

1 *Roads and Railways*

Permafrost can cause problems in all stages of construction from the location of the route to the final surfacing of the road or laying of rails. Changes in permafrost conditions initiated by construction can continue during operation of the road or railway and cause long-term maintenance problems. Roads and railroads differ from other engineering structures (except power transmission lines) because they extend over hundreds of miles and it is possible for permafrost conditions to change along the route thus necessitating variations in construction and maintenance techniques. A number of references are available on this subject including several papers from which the following material was obtained (Canada 1957, Legget and Dickens 1959, Legget and Pihlainen 1960, Nichols and Yehle 1961, Raison 1959).

The selection of the route is a most important element in permafrost areas. Great attention should be given to soils and drainage conditions and the shortest route may not necessarily be the best. Subgrade soil is probably the most important item and fine-grained soils susceptible to intensive frost action should be avoided wherever possible. The availability of sufficient quantities of suitable fill is critical. Cuts in perennially frozen ground should be avoided because excessive soil slumping and icing may occur.

Construction techniques in the continuous zone, where preservation of the permafrost is the prime consideration, differ considerably from the discontinuous zone where thawing will occur or avoidance of permafrost areas may be possible. In the continuous zone the insulation method of construction is relied on to retain the thermal regime because ventilation, as used under buildings to reduce heat flow downward, cannot be applied. In the discontinuous zone the permafrost may be removed by stripping the surface vegetation. Thawing is allowed to proceed naturally or is accelerated by removing soil as it thaws. In some cases, soils may be excavated while they are still frozen.

Construction equipment is often moved to the site in winter when the ground is frozen providing maximum mobility. In the continuous zone the right-of-way should be put through with as little disturbance to the ground as possible. Ideally, this should be done by hand clearing in winter when the ground is frozen to the surface but it may have to be done in summer. The success of the whole job is greatly dependent upon the way in which construction operations are carried out since any unnecessary disturbance of the surface cover will introduce serious problems both during construction and after, due to increased thawing of the permafrost.

Insulation is accomplished by placing a protective layer of gravel directly on the undisturbed ground. The thickness of ground required to preserve the perma-frost will vary with many local factors. It may prove impracticable to provide sufficient thickness to prevent any thawing below the fills. Roads have been built over permafrost using only two to three foot gravel fills. Borrow pits have to be opened sufficiently in advance of construction to allow thawing of the material to be used for fill. Granular coarse-grained materials may not always be available necessitating the use of whatever materials occur locally.

In cases, both in the continuous and discontinuous zones, where frost-suscep-tible soils are encountered and large ice content exists, insulation alone is not sufficient and it may be necessary to excavate the frost-active soil and refill with granular material. This same method of excavation and refill is often used in hilly areas where cuts have to be made as well as fills. The excavation must be per-formed quickly and the fill placed immediately to prevent the permafrost from thawing below the excavation. The fill should be deep enough to stabilize the permafrost table in the zone of the fill. In all cases where fill is used, some settle-ment can be expected during the first year due to the compaction of the subgrade. Stage construction and delaying final surfacing of the road may be desirable until a new permafrost regime has been established.

Cuts and fills are very critical in road construction on permafrost. A general rule is to use fill and avoid cut sections where possible. Cuts entail the removal of part or all of the seasonally thawed layer and perhaps some perennially frozen ground. This can lead to loss of bearing capacity, interference in subgrade drain-age, and the initiation of landslides and also of surface icing, especially in silt soils with high ice contents. If cuts are unavoidable, the unstable material should be replaced with coarse-grained material or the road should be relocated.

During the first winter following the placing of the fill, the permafrost table rises into the roadbed and parallels the general cross-section of the fill itself. During the following spring and summer the upper thawed layer of material may slide and creep on the surface of the frozen layer beneath. A problem can arise in road sections running east-west where the depth of thaw is greater on the south side of the road – this results in soil movement on that side of the road. Insulating peat cover on the south side can help to remedy this problem and maintain equal thaw penetration on both sides of the road.

There are no paved roads in Canada's permafrost region, although the Alaskan section of the Alaska Highway is paved. Concrete surfaces perform satisfactorily where the permafrost table is at least nine feet below the limit of seasonal freezing and thawing or where the foundation soils are not frost susceptible. Where the road passes over silty soils, the surfacing operation should be done late in the year after the seasonally thawed layer has frozen. An insulating layer of brush or timber is laid on the roadway, then this is overlaid by a gravel mat having a minimum thickness of four feet. The final surfacing should be delayed until the new permafrost regime has been established, i.e., until the permafrost table has readjusted to the new thermal regime created by the road.

Adequate drainage of roads is essential because even very cold water will thaw the underlying permafrost. Drainage is a particularly serious problem if the general flow of water in the area is perpendicular to the road direction. Surface water is led away in ditches. Deep frost penetration in winter can be prevented in these ditches by filling them with snow.

If the seasonal frost does not penetrate to the permafrost in winter, ground-water in the intermediate unfrozen layer can cause problems. The repeated application of wheel loads on a fill over saturated subgrades of fine-grained soils causes a puddling and pumping action which tends to reduce the stability of the subgrade and bring water and subgrade material upwards into the fill. Frost boils may cause roads to become impassable in the spring just as in temperate regions.

The most difficult of all materials to drain is peat which extends over large expanses of Canada's permafrost region. It is usually waterlogged because it can hold up to 1,500 per cent moisture by weight. It is impossible to drain by ordinary methods and it often obtains what little strength it has only from the surface mat of partially dried organic matter.

Icings, which are associated with drainage, occur frequently on roads in permafrost regions and adequate protective measures against these features are vital. This account of the cause and control of icings is derived mainly from the paper by Thomson (1966) based on his experiences on the Alaska Highway. The obstructions of surface or underground natural drainage courses in winter causes water to accumulate on the surface and freeze, forming large sheets of ice which can block a road completely. Icings may be ten feet thick and one-half mile long.

There are three types of icing. One type develops from ground water lying in the thawed ground above the permafrost on the uphill side of a road. In winter the frost penetrates to the permafrost in the cleared roadway forming a barrier to downslope groundwater movement. As the seasonal frost penetrates towards the permafrost above the road, this water is forced to the surface where it freezes quickly filling ditches and spreads over the road if not controlled. The second type occurs where water issues from a perennial spring and freezes on the surface of the road. The third type is formed where streams or rivers freeze to the bottom and water is forced out of the bed building up large masses of ice which may endanger roads and bridges.

Maintenance problems arising from icings fall into two broad categories. The first is the hazard to driving. The ice surface is slippery especially if covered with a film of water, and it is frequently hummocky and rough to drive over; the road may even become impassable if ice accumulates in sufficient quantities. The second problem is road damage which may take two forms. Small bridges and culverts may be distorted by the ice; and melting ice water softens the roads in the spring.

The primary goal of control and correction is to keep the road surface free of ice. Passive measures used in winter to remove the ice after it has formed are costly and laborious. A mobile steamer may be used to open narrow slits in ice-filled culverts but visits are required almost daily in subzero weather. Ice may be dynamited as it accumulates and the fragments cleared by grader or bulldozer but this is also expensive. Dynamite may also be used to open seepage channels in the ice for the winter. Ice may also be scraped off the road by a grader often equipped with a scarifier attachment. This treatment is required repeatedly to be effective. A final passive measure is the use of a fire pot, a 45-gallon drum in which a fire is kept burning most of the winter, placed at the mouth of a culvert. This usually keeps the culvert open and prevents ice from encroaching on the road. Fire pots can only be effective for very small areas and must be visited frequently.

The so-called active measures are taken during the summer to divert the water away from the road and are more practical than passive measures. Good drainage, which is vital to icing correction can be effected by surface trenches or subsurface drains, both constructed during the summer. Freezing belts are areas of ground, upslope from the road which are stripped of vegetation and kept clear of snow until about January to ensure maximum frost penetration. The icings form here and do not reach the road. To be successful there must be a layer of permafrost beneath that can be intercepted by the downward winter freezing. The vegetation must be replaced or some other insulation provided in the summer to protect the permafrost from thawing. Water may be intercepted away from the road and ponded behind earth dykes. An alternative is sackcloth stretched on stakes across the seepage course. When the sackcloth becomes wet, it freezes and ice accumulates behind it, and then when the ice reaches the top of the stakes, new stakes are installed and the process repeated. Another method occasionally used is simply to raise the road grade above the normal icing level. Installation of oversize culverts placed on steep gradients and as deep in the ground as possible will help in icing control. Sackcloth hung over both ends of culverts often prevents ice from forming in them.

In the case of river icings, some of the same methods are used but with modification. To intercept river water, freezing belts have to stretch across the valley and extend deeper into the ground. In a braided stream the channel may be deepened annually to increase current velocity and decrease the danger of rapid channel freezing.

2 Airfields

The great area required for an airfield makes the location of a suitable site a difficult problem in permafrost areas (Legget and Dickens 1959, Linell 1957, Sebastyan 1963). The centre line of a runway is governed by the usual factor of wind, unobstructed air approaches, drainage, subgrade, and amount of grading. On permafrost more weight has to be given to surface drainage, subsurface drainage and subsoils. Fine-grained soils should be particularly avoided for airfields. Wherever possible, the runway should be located on coarse-grained soils; gravel eskers are particularly suitable.

As with roads and railroads, preservation of the permafrost is the prime consideration in the continuous zone; thawing will occur in the discontinuous zone or it may be possible to avoid permafrost areas. In the continuous zone the insulation method is used to protect the permafrost from thawing and granular coarse-grained insulating blankets eight feet and more thick have been placed on the ground surface. White or light-coloured finished surfaces are preferred to black to reduce surface temperatures which can be of particular significance in the thermal performance of the airstrip. In the discontinuous zone the permafrost may be handled in the same way as was described previously in the section on roads.

The treatment of fine-grained soils which are susceptible to frost, and drainage problems are very important considerations. It is desirable that the centre line of runways be laid parallel to the movement of surface and subsurface water because subsurface drains will not work in winter months. Side drainage ditches should be constructed far enough from runways to leave room for snow piling.

With the development of increasingly heavy and high speed aircraft, the final finished surfaces of runways must be smooth so as not to interfere with taxiing, take-off, and landing operations. Differential settlement and excessive cracking cannot be tolerated. This means that the underlying permafrost in the continuous zone must be preserved and measures taken to counteract frost action.

3 Bridges

Unlike road and railroad construction, which varies depending on whether the structure is in the discontinuous or continuous permafrost zone, specifications for bridge construction must, in all cases, be based on the bearing strength of the soil in its unfrozen state. This is because the effect of running water in the valley to be traversed is either to lower the permafrost table considerably or to cause the disappearance of the frozen condition of the ground (Canada 1957).

Massive concrete piers are recommended for bridge foundations in permafrost areas. In areas where permafrost is widespread, although depressed below the river, massive poured abutments should be placed at least 3½ feet in the perennially frozen soil, if it is coarse-grained, and at least 5 feet in fine-grained soils which are frost susceptible. Pile foundations are preferable for medium to small span bridges where the depth of thaw exceeds 6½ feet or the permafrost is degrading and the embankment exceeds 16 feet in height. Exposed pier facings should be reflective to the sun (Canada 1957). Frost heaving is a critical problem particularly at abutments and pile locations in shallow water.

Icings can damage bridges extensively. They are most troublesome in broad, shallow, gravelly channels with steep gradients where the stream freezes to the bottom and frost penetrates deeply into the gravel forcing extra water to flow above the ice. This freezes and builds up ice accumulations. River icings seldom occur in streams of low velocity with deep narrow channels and overhanging vegetative growth along the banks (Canada 1957).

4 Pipelines

With the expansion of oil exploration activities in northern Canada and the discovery of oil in northern Alaska, speculation on the most satisfactory methods of constructing and operating pipelines has increased in recent years. Various possibilities are being considered but it is not yet known which will prove technically feasible and the most economical (Harwood 1969). Basically, pipelines can be placed either in the ground, on the ground surface, or suspended above ground. These variations all have advantages and disadvantages.

One consideration is the distribution of permafrost. In the discontinuous zone, permafrost is patchy and its temperature is close to 32° F. In the continuous zone, permafrost occurs everywhere beneath the ground surface and its temperature is several degrees below 32° F. Detailed site investigations will be required along the proposed pipeline route to assess the soils and permafrost conditions. This is particularly vital in the discontinuous zone where the pipeline will pass through both permafrost and non-permafrost areas. Here, it will be necessary to design for the properties of the soils in the frozen state and also in the thawed state where permafrost does not occur or where it may thaw as a result of construction and operation of the line.

A pipeline buried in the ground has the advantage of being subjected to a much smaller annual range of temperatures (perhaps 40° F) versus above ground where air temperatures can range from −60° F to +90° F (range of 150° F). These temperature fluctuations have important implications for the viscosity of the oil and the volume changes of the pipe itself due to expansion and contraction. Serious problems arise, however, in coping with the soils and permafrost conditions. A pipeline buried in the active layer would be subjected to annual freezing and thawing of the ground and severe frost heaving in fine-grained soils. If a pipe is laid in the permafrost, the problem arises first of excavating a trench in the frozen ground. Thawing of the permafrost would cause considerable settlement in areas of high ice concentration. These ground movements in both the active layer and the underlying permafrost would pose serious problems in maintaining the ground stability necessary for large diameter pipelines.

A further problem is presented by the oil itself. It comes out of the ground at a temperature exceeding 150° F. If warm oil were pumped into a pipeline buried in the permafrost or the active layer, thawing of the frozen ground could ensue. This problem could possibly be circumvented by insulating the pipe. Another alternative is to cool the oil below 32° F but this would increase the viscosity perhaps to the point where a viscosity breaker would have to be installed at the pumping station. The possibility of wax deposition also increases at these lower temperatures.

A pipeline laid on the ground surface would be subjected to the large annual range of temperature mentioned above with accompanying viscosity and line volume changes. Ground surface movement due to frost heaving would be substantial in areas of fine-grained soils. One proposed method, which is receiving favourable consideration, is to build a road, along the pipeline route. A second nearby embankment parallel to the road would carry the pipe and be sufficiently wide to accommodate construction and maintenance equipment. The pipe could be placed along one side of the embankment and gravel banked up around it. The permafrost table would rise into the embankment making it stable and the gravel would protect the pipe from the large annual fluctuations in air temperature. One problem with this method is to locate the tremendous quantities of gravel and other borrow material required to build the road and pipe embankment.

The third method, that of suspending it above ground involves the construction of trestle-like structures to support the line. This arrangement would accommodate movements of the pipe caused by expansion and contraction through the year. It would also eliminate the problems caused by ground volume changes due to frost heaving, but the large temperature fluctuations would impose the problems caused by viscosity of the oil at low temperatures. Effective insulation of the suspended line would be difficult. This method would eliminate the problems caused by ground volume changes due to frost action. However, the supporting structures would have to be very stable, perhaps installed on piles embedded in the permafrost, and spaced to carry the heavy large diameter pipe.

These are some of the aspects which will require consideration in the construction and operation of pipelines in permafrost regions. It is a very complicated multi-faceted problem which will require many interrelated investigations. The method actually employed will depend on cost and technical feasibility which are related to the soils and permafrost conditions encountered along the route, and the properties of the oil itself.

B   EXPERIENCE IN NORTHERN CANADA

1 *Roads*

a / Northwest Highway System
The first major road constructed in Canada's permafrost region was the Alaska Highway extending from Dawson Creek, BC, to Fairbanks, Alaska now part of the road network in the Yukon Territory and northern British Columbia known as the Northwest Highway System. It was built as a military supply route to connect Alaska to the continental United States and to serve a number of airstrips constructed in 1941 in northwest Canada and Alaska. Highway construction began in March, 1942 and a 1,670-mile road suitable for trucking was completed in November of the same year. This pioneer road was mainly 22 to 24 feet wide, but only 18 feet in places and was gravel surfaced (Richardson 1943). In 1944, relocations and cut-offs reduced the original route by nearly 100 miles to 1,585 miles. An access highway from Haines in the Alaskan Panhandle at tidewater near the head of the Inside Passage was built to meet the main highway 108 miles west of Whitehorse (Richardson 1944*b*).

Since the end of the war, branch roads have been added in Yukon Territory and northern British Columbia (as is shown in Figure 30) and considerable relocation and reconstruction of the original Alaska Highway has continued through the 1950s and 1960s. The Alaska Highway itself from Dawson Creek, BC, to the Yukon-Alaska boundary, except the first 83 miles and the Haines Highway, are maintained by the Canadian Department of Public Works who have head-quarters in Whitehorse. The branch roads to Carcross, Dawson, Mayo, and Ross River, and the Dempster Highway being constructed to Fort McPherson, NWT, are maintained by the Canadian Department of Indian Affairs and Northern Development. The Stewart-Cassiar Highway and the road to Atlin in northwestern British Columbia, and the paved 83-mile section of the Alaska Highway from Dawson Creek are maintained by the British Columbia Department of Highways.

The construction of a highway to Alaska had been discussed for years but was not undertaken until the war made it a necessity. Because of the urgent require-ment to complete it the most direct route was followed. No consideration was given to permafrost conditions, and engineering techniques used in temperate regions were employed – resulting in road failures and severe icings at many locations. The following account of permafrost and icings on the Northwest High-way System includes both wartime experiences on the original Alaska Highway and postwar experiences on the whole highway network. Trouble still occurs at some locations but improved construction and maintenance techniques have greatly reduced the difficulties.

The first known patch of permafrost encountered on the Alaska Highway occurs in a peat bog 94 miles north of Dawson Creek about 11 miles beyond the paved section (Brown 1968). The most southerly permafrost of sufficient extent and thickness to affect road construction and maintenance occurs in extensive peat-lands 250 miles north of Dawson Creek. For a distance of 30 miles the road winds through and between peat bogs in which permafrost exists. North of this area, there are only a few widely scattered islands of permafrost near the road, occur-ring in peat bogs. A severe permafrost situation was encountered at Mile 398.4 where the road crosses the side of a hill, when, during initial construction, a road cut was made into the hillside. The underlying permafrost began to thaw causing the road embankment to slip sideways down the slope. The problem was corrected, at considerable expense, by building the road further downslope without a road cut. Permafrost occurs on north-facing slopes in the vicinity of the continental divide at about Mile 481 where the highway reaches its highest elevation at Sum-mit Lake, 4,212 feet above sea level. The highway is, however, virtually un-affected. From here to Whitehorse at Mile 918, permafrost occurs in scattered islands and only small portions of the highway have been built on it.

Between Summit Lake and Watson Lake the highway crosses the Liard River twice. Both river crossings could have been avoided initially by using the south bank route. Use of this route was impractical, however, because of permafrost on the north-facing slopes, poor drainage, and the necessity of numerous rock cuts. The north bank is gently rolling with dry gravel soil, and there is no permafrost

TABLE 3
Percentage of Alaska Highway Built on Permafrost between Continental Divide and
Whitehorse (Eager and Pryor, 1945)

Highway section	Length (miles)	On permafrost (miles)	Percentage
Prochniak Creek to Lower Crossing (Liard River)	16.4	0	0
Lower Crossing to Upper Crossing (Liard River)	146.2	0.5	0.3
Upper Crossing to Nisutlin Bay	160.9	7.0	4.4
Nisutlin Bay to Teslin River	33.1	9.0	27.2
Teslin River to Whitehorse	80.3	2.0	2.5
TOTAL	436.9	18.5	4.2

in the south-facing slopes. Finally the latter route was chosen despite the necessity of constructing two large bridges. It was estimated that the adverse soils and
permafrost conditions on the south bank route would create higher construction
and maintenance costs (Sturdevant 1943).

Table 3 shows the number of miles and percentage of the original Alaska
Highway built on permafrost from the continental divide to Whitehorse.

Northwest of Whitehorse, about 28 per cent of the highway was built on permafrost in contrast to the 4.2 per cent southeast of the city (Denny 1952). Virtually
no permafrost was encountered from Whitehorse to Mile 1,113, about 200 miles
to the northwest. From here to the Alaska boundary, permafrost is widespread
in mineral soil as well as peat bogs. The most difficult construction problems
caused by permafrost were encountered northwest of Whitehorse although the
phenomenon caused problems southeast of the city too. During the initial construction engineers, with no previous experience in permafrost areas, tried to
ditch a considerable stretch of the highway and encountered difficulties. In some
places the road sank 10 to 15 feet as the permafrost was thawed by heat penetration through the ditches. The permafrost table lay only 1 to 2 feet beneath the
ground surface in the undisturbed areas and the right of way was easy to clear
because of the shallow tree roots. After clearing, the topsoil was stripped and
attempts were made to grade the road. As fast as the insulating vegetation cover
was removed, the ground thawed, creating a morass that became worse the more
it was worked. As a remedy, the vegetation was left in place and a layer of brush
and timbers 4 to 5 feet thick was added. Gravel fill was then placed on top by end
dumping. The biggest difficulty in this construction was finding enough unfrozen
material. Fortunately, gravel was plentiful and it did not have to be hauled more
than 18 miles (Richardson 1943).

The most severe problems were encountered along a 90-mile stretch between

Edith Creek in western Yukon Territory and the Alaska boundary where more than half of this length of road had to be closed during the summer because of extensive slumping caused by thawing. The poorly drained material could not be removed and there was not enough unfrozen fill to dump over the brush and timber mat placed on the frozen ground. Therefore, the only thing that could be done was to clear the trees and brush and place a shallow surface fill of whatever frozen gravel was available. The frozen base was not sufficiently insulated, however, and severe damage from thawing ensued (Richardson 1944).

At the north approach of the bridge crossing the Donjek River in western Yukon Territory, a cut was extended thirty feet into perennially frozen ground. Four to five feet of gravel was placed as subgrade over this cut and the road surface has performed satisfactorily, although the surrounding side hill has thawed and slumped to some extent. While it was anticipated that this approach would have to be refilled periodically for perhaps three or four years, apparently the correct amount of insulation was placed to retain the perennially frozen state at the bottom of the cut. Difficulties have occurred at bridge approaches where the overburden is eroded by flash floods causing the permafrost to thaw and the bridges to settle. These problems are being rectified as the original temporary bridges are replaced (Love 1954).

Some of the most difficult highway construction and maintenance problems have occurred in areas of perennially frozen peat. If the peat is less than ten feet thick, it is removed and replaced with coarse-grained fill. The usual practice is to scrape off the vegetative cover and remove the peat in layers as it thaws down over a period of several weeks. This is a slow process but more effective than blasting. Because of its high ice content and elasticity, frozen peat absorbs much of the explosive energy and usually breaks into large unmanageable chunks (Thompson 1957).

If the frozen peat is more than ten feet thick, excavation is too costly and time-consuming. The usual method is to place gravel on the ground each winter for several years until the fill is sufficiently thick and wide to be used as a road; during this period, the permafrost table rises into the fill and is maintained at a high level by keeping the road clear of snow and placing peat insulation on the shoulders. This method spreads construction costs over several years but a large quantity of fill is required. If the fill is frozen or has to be hauled long distances, then road costs will be very high.

Small patches of permafrost were encountered at one location on each of the branch roads in northern British Columbia – the road to Altin, and the Stewart-Cassiar Highway. They were of minor concern and have apparently not caused any difficulties since construction.

Permafrost is generally widespread on the roads maintained by the Department of Indian Affairs and Northern Development in the Yukon Territory because they extend mostly north of the main Alaska Highway. Great care has to be exercised in construction and maintenance particularly on hillsides where disturbance by road cuts can initiate thawing of the permafrost and severe soil slumping. The most northern stretch of this road network is the Dempster Highway, extending 75 miles northeast of Dawson in the north central Yukon Territory (Figure 30). It will in the future be extended to the northeast an additional 260 miles to Fort McPherson, NWT. Construction began in 1958 and the first 75-mile section was completed in the early 1960s. It is located in the northern portion of the discontinuous permafrost zone; the thickness of permafrost varies from about 200 feet in the vicinity of Dawson to perhaps 1,000 on the higher plateaus to the north. The northern section of the finished road near Fort McPherson will lie in the continuous zone where permafrost may be considerably thicker than 1,000 feet.

Clearing along the first fifty miles of the proposed road was carried out in the fall of 1959. Areas of perennially frozen ground were marked off, to be left undisturbed by machinery, and the cuttings piled in the centre of the right-of-way. This layer of trees and brush, piled on the undisturbed moss-covered ground surface, acted as additional insulation helping to maintain the frozen condition of the ground. Areas with no permafrost were cleared and the brush burned (Trueman 1962).

The initial planning of the Dempster Highway was based on modern highway standards and practice. The widespread distribution of permafrost, however, resulted in a scarcity of available materials for road construction, and reduced standards of grades and curvature were found to be necessary. Route selection was placed on a more flexible basis following, where necessary, the line of material availability rather than the original survey. On this basis a contract was awarded in 1960 and construction of the route closely followed the North Klondike River where gravel bars provided a reliable source of material. Farther north the lack of gravel forced the route to follow the banks of the Blackstone River.

By 1961 considerable work was required on the constructed road due to thaw settlement and icing conditions. In June, a contract was awarded for repair work on this section and for the completion of new construction to Chapman Lake about twenty miles north. The new road followed the west bank of the Blackstone River, and was completed to four miles beyond the original objective. Special care had to be given to the protection of the permafrost to keep thawing to a minimum. Side hill construction on frost-susceptible soils was avoided wherever possible. Stretches of road, built over frost-susceptible materials on relatively level ground, were elevated two to three feet above the surrounding terrain. Frost heaving in the subgrade during the first year or two before the permafrost rose into the fill made the road surface very uneven necessitating regrading as part of the maintenance programme.

During the maintenance work it was noted that the granular borrow pits that had been stripped previously to the permafrost table had thawed to a depth of two to three feet during the summer season. This provided an economical source of material for the maintenance programme. Measures had to be taken also to minimize the effects of icings – to prevent icings on the road, a ditch and dyke were constructed up slope from the road, forcing groundwater to the surface and forming the icing at a point where no damage occurred to the road.

As mentioned previously icings are prevalent and troublesome (Thomson 1957 and 1966). During the winter of 1955–6, for example, more than $50,000 was spent in combating them. Individual icings cost from about one or two hundred to several thousand dollars per year depending on their size, location, and distance from the maintenance camp. Between Dawson Creek and Whitehorse, 92 major and 57 minor icings have been encountered and from Whitehorse to Big Delta, Alaska, 100 miles southeast of Fairbanks, 34 major and 38 minor ones. More than half of them occurred where natural conditions would have produced some icing if left undisturbed but the amount of icing was greatly increased by road construction. Seepages and small springs caused most of them (Eager and Pryor 1945).

Although the presence of permafrost is not essential for icing to occur, it has been observed along the Alaska Highway that most icings were encountered where permafrost existed. For example, between Whitehorse and Big Delta, 68 per cent of all the major icings occurred where road construction took place on permafrost (Eager and Pryor 1945). One large icing, which has never been counteracted successfully, occurs on Haines Road just south of Haines Junction. Here a sheet of ice, fed by groundwater from a nearby slope, has at times covered one-half mile of road to a depth of eight to twelve feet. From February 6 to March 21, 1957, $10,000 was spent in clearing the ice by bulldozer. Most of the icings southeast of Whitehorse occur on a 150-mile stretch of road between Nisutlin Lake and Teslin River where islands of permafrost are numerous.

b / Canol Road

Another road constructed during the Second World War was the Canol Road extending 620 miles southwest from the west bank of Mackenzie River opposite Norman Wells, to Whitehorse. A 4-inch pipeline to carry crude oil from Norman Wells to a refinery in Whitehorse ran alongside the road. At Norman Wells a tank farm was constructed and pumping stations were erected on the pipeline at 50-mile intervals. Hauling of equipment down Mackenzie River having begun in 1942, the project was completed in April, 1944. The road had to be built at great cost over the Mackenzie Mountains and could be used only by vehicles having four-wheel drive owing to steep grades. Except at the southern end, the road was entirely constructed on permafrost as follows: brush and trees cut from the right-of-way were piled on the centre of the road; side ditches were then cut down to the permafrost table and the excavated soil placed over the brush and trees to form a base for the road fill; and coarse-grained material was then hauled to complete the grade (Richardson 1944a). The road performed satis-factorily, except for some stretches of icings, during its short period of use from April, 1944, to May, 1945. At the end of the war the road was closed because Norman Wells oil could not compete in southwest Yukon Territory with oil brought from the south. Although the use of the Canol road terminated because of economic factors, permafrost was instrumental in increasing construction difficulties already caused by remoteness and rugged terrain.

c / Mackenzie Highway

The first road constructed into the permafrost region after the Second World War was the Mackenzie Highway, built from Grimshaw in Northern Alberta to Hay River, NWT, on the south shore of Great Slave Lake, on which construction was begun in October, 1946 and completed in 1948. As part of its Roads to Resources programme in Northern Canada, the federal government in 1960 completed the highway 280 miles around the north end of Great Slave Lake to Yellowknife. The only uncompleted link in this new section is a bridge across Mackenzine River near Fort Providence. A ferry transports vehicles across the river in summer and they cross on the ice in winter. In the early 1960s the road from Grimshaw to Hay River was rebuilt to the same standards as the extension to Yellowknife.

Now the most southerly occurrence of permafrost on the Mackenzie Highway is found about 120 miles north of Grimshaw, in the vicinity of Keg River. North-ward to Hay River it occurs mostly as scattered patches and islands in peat bogs, ranging in thickness from a few inches or feet in the south to about fifty feet at Hay River. Permafrost has been encountered also on some north-facing slopes of east-west oriented river valleys such as the bridge crossing at Meander River in Northern Alberta. From Hay River to Yellowknife, the road extends through widespread discontinuous permafrost which extends to depths of 150 feet at Yellowknife.

From Grimshaw to the Northwest Territories boundary, construction costs of the Mackenzie Highway were shared equally by the Province of Alberta and the federal governments. Little permafrost was encountered along this section and the design of the highway was not influenced by this factor. North of the boundary, permafrost is more widespread and the federal government, which assumed the

total cost of construction, found that modifications in highway design were necessary to cope with this phenomenon (Wallace 1961; Savage 1965).

Special clearing of the right-of-way, consisting of cutting trees and brush within
one foot of the ground surface whether in permafrost areas or not, was specified.
Permafrost areas were designated on the plans as "special clearing areas" and
clearing was performed so that the organic mat of mosses and peat was not
damaged. The cleared material was placed in a flattened layer over the embankment area to improve the insulation of the perennially frozen soil against thawing
and thereby maintain it in its frozen state. In the non-permafrost sections, the
cleared material was either burned or buried, or in some cases, such as low-lying
poorly drained peat areas, it was placed in the embankment.

Drainage facilities were given careful consideration to prevent thawing of the
permafrost by standing and running water. Ditching was done immediately after
clearing, and before any other earthwork was carried out, so that the soil beneath
the embankment would be as low in moisture content as possible while the embankment was being placed. Additional water was squeezed from the underlying
soil during placing of the embankment but it was carried away in the ditches. Thus,
the possibility of the permafrost thawing because of water lying adjacent to the
grade was mostly eliminated. Berms were left between the ditch and the toe of the
embankment, in contrast to normal procedures where ditches are placed adjacent
to embankments, after which the permafrost table in the vicinity of the ditches
receded because of thawing. This was of no concern, however, because the soft
wet zone caused by the thawing permafrost was not near the embankment.

The installation of culverts presented special problems. They could not be
placed on the existing ground surface because water flowing through would thaw
the underlying permafrost. The best solution was to excavate to some depth and
backfill with gravel quickly before any thawing began. The coarse-grained material usually provided sufficient insulation to reduce thawing to a minimum. In
fact if the gravel extended below the permafrost table, it appeared that the permafrost table would rise into it after the pipe and embankment were placed. On the
Yellowknife portion of the highway north of Hay River where permafrost is
widespread, the bottom of some culvert excavations was lined with moss and
peat before placing the gravel. This was done in an attempt to restore the natural
insulation properties of the organic cover and it appears to have been successful.
Special culvert systems were used to counteract icing and flash flood conditions
which could cause serious erosion in perennially frozen (especially with fine-
grained) soils. On the Mackenzie Highway south of Hay River, dual-staggered
installations of culverts were used where icing was anticipated. These installations
were in relatively high earth fills. A pipe was placed in the lower portion of the
fill to handle normal summer flow and some icing of this structure was acceptable.

A second pipe was installed in the fill above and to one side of the first pipe. This second pipe does not normally carry any flow during the summer or early winter, so little or no icing occurs and its full capacity is available for spring flooding conditions.

Permafrost was encountered occasionally in borrow pits but where possible this material was not used for road embankments. Where frozen material was the only material available, the top thawed layer was removed allowing a thin layer beneath to thaw (this was subsequently removed). The material had to be spread in the proposed embankment in thin layers and allowed to dry to approximately optimum moisture content before compaction and placing of another layer could be undertaken.

d / Thompson Highway

Construction on this road, connecting the nickel mining town in northern Manitoba to the province's road system, was begun in the early 1960s and completed in 1965. Along the entire route from the southwest end, where it joins the Snow Lake-Wekusko road, to the northeast end at Thompson permafrost occurs in scattered islands varying in thickness from about 5 to 50 feet. The first 40 miles at the southwest end pass through the Great Wekusko Bog where the terrain consists of low-lying wet depressions filled with peat 9 feet deep interspersed with extensive peat plateaus containing permafrost. A special experimental design was formulated by Manitoba highway engineers to cope with the permafrost and peat bog conditions (Newman 1963; Construction World 1963).

The usual method is to carry out construction in summer, removing all wet peat from the depressions and blasting out perennially frozen peat plateaus on the right of way, and hauling in fill over long distances. Instead construction was carried out in winter. After trees and brush were cleared from the right of way, the snow cover was removed. Tractors and snowmobiles moved over the depressions to consolidate the peat. This process was first done with light vehicles, then repeated several times with progressively heavier vehicles. In the lower winter air temperatures of $-30°$ F, the top layer of compacted peat froze solidly to the same strength as the adjacent perennially frozen peat plateaus. This frozen foundation was then covered with a 4-foot layer of impervious silty clay, which when compacted and trimmed provided a 3-foot cover for the permafrost base. This clay blanket sheds summer rain and protects the underlying permafrost from thawing. A layer of gravel was placed on the clay to provide the finished road surface. Ditches were placed a minimum of 85 feet from the shoulders so that water would not affect the permafrost beneath the road. Thermocouple cables have been placed at several locations along this section of the highway to measure ground temperatures. Observations indicate that the base is remaining frozen and retaining its strength. This experimental utilization of permafrost as a road foundation in the North is being watched closely and may be used widely if proven successful.

e / Hanson Lake Road
Opened in 1963 this highway begins at the small town of Smeaton, east of Prince Albert, and extends 240 miles to the northeast terminating at Flin Flon, Manitoba. The eastern portion lies just within the southern fringe of the permafrost region. The first 20 miles at the southwestern end is poplar wooded and level, with some peat bogs, followed by 50 miles of jackpine and sand, also with some peat bogs; then for the next 30 miles the route is through high hills and numerous peat bogs where there are islands of permafrost. The soil here is predominantly heavy clay and construction is difficult; in the next 50 miles there is some sand, limestone outcrop, and extensive peat bogs with islands of permafrost; and the last 90 miles is through rugged Precambrian rock hills and ridges up to 400 feet high with some peat bogs and permafrost in depressions (Roads and Engineering Construction 1959).

Permafrost occurs in northern Saskatchewan in varying degrees of extent and thickness and it caused some difficulty on the Hanson Lake Road. Because this road is located in the southern fringe of the permafrost region, the occurrences of permafrost are patchy and thin. Thus, where permafrost was encountered, such as in the area mentioned above, the surface vegetation was stripped to allow thawing and consolidation of the frozen ground. Care was necessary to ascertain that all permafrost was thawed at any location before the road grade was built to prevent settlement. On one section, permafrost ten feet thick was encountered. Here the overlying moss cover was stripped and the frozen soil cleared in layers as it thawed. Four months of thawing were required before enough soil could be removed to permit construction of the grade. The entire thickness of permafrost was not removed, however, and additional settlement due to thawing took place the following summer.

f / Roads in settlements
In addition to the highways extending many miles across Canada's permafrost region, at many settlements there are small road networks to carry local traffic. This traffic is limited becase of the lack of connection between settlements, but construction methods and problems imposed by permafrost are similar to those of the highways already discussed. The main difference is that soils and permafrost conditions throughout any one settlement site are generally fairly uniform and more or less the same construction problems and techniques prevail anywhere in contrast to highways where soils and permafrost conditions change along the route.

124

At Uranium City, on the north shore of Lake Athabaska, permafrost occurs in scattered islands. Along some stretches of road laid on permafrost, thawed saturated soil flowed out from the sides resulting in considerable settlement (see Figure 31). In some areas, as much as 17 feet of road fill was placed and subsequently submerged, causing high maintenance costs.

At Thompson, Manitoba, permafrost occurs in scattered islands 5 to 50 feet thick in varved clays. In the construction of roads particular attention was paid to the removal of all organic material, which varied greatly in depth. Sub-cutting was considered unnecessary because topping up after subsidence was used extensively. Road building aggregate was readily available so that both rock and sand mats were utilized to overcome poor subgrade zones. The application of pavement, consisting of normal asphaltic concrete was usually delayed for at least one year, and, circumstances permitting, two years, in order to allow for further melting of ice layers and subsequent subsidence. The maintenance of roads involves raising and reconstruction of subsided curbs, sidewalks, and paved surfaces. Differential settlement caused by thawing of the permafrost has reached a maximum of 4 feet and affected about 5 per cent of the roadways. The raising of

FIGURE 31 Perennially frozen silt with high ice content has thawed and is flowing out from roadbed at Uranium City, Sask., in discontinuous permafrost zone. Very unstable conditions result from this phenomenon.

curbs without damage to curb sections is most difficult – various methods of curb capping and pressure grouting have been tried, with little success. The filling of low areas in the pavement surfaces with fresh asphaltic concrete may prove to be the solution to this problem. Some problems being experienced with road surfaces, curbs and walks could have been decreased greatly if this phase of construction had been delayed five years or more until thaw settlement of the perennially frozen soils ceased. Under the circumstances, however, it was decided that the advantage of immediate paving and servicing of much of the area outweighed the disadvantages (Klassen 1965).

In the Mackenzie River valley, permafrost problems have been encountered during construction and maintenance of roads at Norman Wells and Inuvik. Vehicle traffic in these two towns is heavier than in any other settlements on the Mackenzie River, resulting in greater maintenance problems.

At Norman Wells, where permafrost is discontinuous but widespread and about 200 feet thick, stripping the moss cover resulted in thawing and such unstable conditions that it appeared impossible to construct any type of satisfactory permanent road. The solution lay in not disturbing the existing moss cover under which the permafrost table was near the ground surface. Additional moss and brush from the sides of the right of way were packed on top of the natural moss cover to form the road base and heavy clay or gravel subgrade from borrow pits was placed on top. In this non-frost-susceptible grade the permafrost table was held at a high level and even rose into the grade after several years. Because some of the most easily obtainable soils in the Norman Wells area are frost-susceptible silts, frost boils occurred on the roads where this was used. Drainage improved conditions to some extent but did not eliminate the problem entirely. During the spring break-up heavy traffice has to be kept to a minimum because the roads are soft owing to melting ice in the sub-base (Hemstock 1953).

During the winter, a road was maintained across the frozen Mackenzie River to Camp Canol, five miles west of the river on the old Canol Road. At one location, where the road crossed a creek, severe icings occurred, building to thicknesses of 10 to 15 feet (see Figure 32). Control was effected by supplying sufficient heat from oil stoves to the water so that it could be led away from the road before it froze. This was very expensive and only economically feasible because of the availability of cheap fuel at Norman Wells. Portable steam generators were used to control small icings and tractor-drawn rooters and ploughs were used to break up and remove ice from the road surface. This method of control proved costly and heavy machinery was needed during the coldest part of the winter when maintenance and personnel problems were most difficult (Hemstock 1953).

Road construction began at Inuvik in 1955 and continued for several years during which time a total of nineteen miles of road was constructed – giving this settlement the most extensive road network north of the Arctic Circle. The first step was to place a layer of felled trees and brush on the underlying ground surface to protect the underlying permafrost from thawing. Where the ground surface was inadvertently disturbed, the permafrost thawed and water from melted ground ice accumulated beside the road. In one location organic material was added to the brush layer to provide extra insulation for the permafrost. Gravel was obtained from benches near the river which were stripped and allowed to thaw. A minimum of eighteen inches of gravel was placed by backfilling only over the brush and graded by bulldozers to complete the roads.

Most of the roads have performed satisfactorily but some problems with permafrost have arisen. A few very localized subsidences have occurred where underlying massive ice layers melted due to thawing and erosion by water. In some areas, road fills interfered with natural drainage which eroded the grade. Standing water, held by the impervious permafrost, also caused more thawing of the underlying permafrost and subsequent settlement. In one location, a cave-in caused by drainage water erosion, extended ten feet under the road.

FIGURE 32 Icing on road at Norman Wells caused by water issuing from ground between freezing active layer and permafrost, and freezing in cold air temperature conditions.

g / Future construction

Several new roads are planned to enlarge the Northwest Highway System in Yukon Territory. Eventually, a road will encircle Great Slave Lake. The road to Fort Simpson will be completed in 1970 after which construction will begin on a road to Fort Liard and Fort Nelson in Northeastern British Columbia. Farther north, construction is expected to begin in the near future to link Tuktoyaktuk, Inuvik, and Arctic Red River to Fort McPherson and the eventually to be completed Dempster Highway. A road will probably also be constructed from Fort Simpson to Arctic Red River thus connecting the Mackenzie Delta with the south. Additions to existing provincial road networks, built in co-operation with provincial governments, will link northern communities, such as Uranium City, with the south. Careful site surveys will have to be undertaken so that the most direct route, coupled with the least troublesome soils and permafrost conditions may be found.

In the southern reaches of the permafrost region, where its distribution is patchy, as in southwest Yukon Territory and southern parts of the Northwest Territories, it will be possible to avoid most permafrost areas. Although this may result in extra road mileage, the saving in construction and maintenance costs would probably justify it. Farther north in central Yukon Territory and north of Great Slave Lake, some areas of unfrozen ground will be encountered but techniques for construction on permafrost will predominate. In the continuous zone there is no chance of avoiding permafrost and the route of any road, such as the extension of the Dempster Highway to Fort McPherson, will have to be chosen to encounter the least troublesome permafrost conditions.

2 Railways

a / Hudson Bay Railway

This railway, first planned in 1908 but not completed until 3 April 1929, was the first constructed in the permafrost region of Canada (see Figure 33). It was built to provide a shipping route for Prairie grain to Hudson Bay from which the ocean distance to European markets is shorter than from Montreal. Ice conditions in Hudson Strait, however, reduce the navigation season to only about four months in contrast to eight or nine months on the St. Lawrence. This reason, among others, delayed the decision to construct the Hudson Bay Railway for some years (Innis 1930).

In 1908 the railroad from southern Manitoba reached The Pas. The original plan was that the Hudson Bay Railway would be built to Port Nelson, 150 miles south of Churchill. By 1913, 80 miles of track had been laid and by 1917 the rails extended to Mile 332 near Gillam and the grade was completed to Port Nelson a further distance of about 100 miles. Further construction was suspended because of the war and trains operated northward only to Mile 214. At this time it was decided that the shallow harbour at Port Nelson would hinder its development as a port. It was not until 1926 that Churchill was chosen as the terminus and completion of the railroad from Mile 332 was begun the following year (Innis 1930).

The most southerly islands of permafrost on the railroad were found in peat bogs at Wabowden, 54°50′ N. Northward, the islands of permafrost increased in number and size and occurred most frequently in the peat areas. Between Gillam and Churchill permafrost was found everywhere beneath the ground surface. A few miles south of Churchill the railroad crosses the treeline into the tundra which has little relief and elevations varying from 50 to 200 feet above sea level. Numerous shallow lakes and ponds dot the tundra and the land is covered with moss beneath which there is a layer of peat varying in thickness from 4 to 18 feet. Till consisting of grey clay and boulders lies beneath the peat. Ice lenses occur in the peat and till. The ground surface is hummocky and the maximum depth of summer thaws ranges from 12 to 18 inches (Charles 1959).

It was discovered that ten years between 1917 and 1926 of practically no maintenance on the line from The Pas to Port Nelson had resulted in severe deterioration. The spruce ties which had rotted were replaced eventually by 600,000 hard-

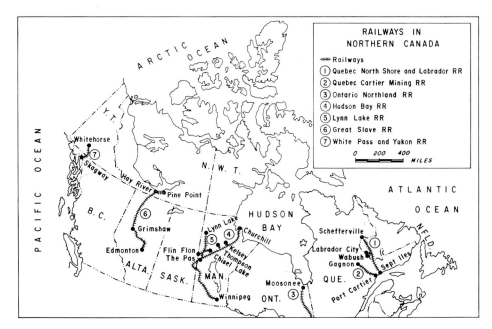

FIGURE 33 Railways in northern Canada.

wood ties shipped from British Columbia. Severe frost heaving had twisted the rails and in some places had pushed them completely off the roadbed. One million cubic yards of new coarse-grained fill was needed to restore the bed itself. Subsidence from thawing was severe in many places. Telegraph poles were heaved out of the ground by frost action. Because the permafrost table lay only one or two feet beneath the surface it was impossible to obtain fill with steam shovels or mechanical grading equipment. Eventually the whole bed was rebuilt with pick, shovel, and wheelbarrow (Fleming 1957).

To relocate the railway from its second crossing of the Nelson River to Churchill, the most economical route was to follow the Port Nelson line to Mile 356 and then due north for 150 miles to Churchill (Fleming 1957). The route was surveyed in the winter of 1927–8 and construction began the following summer. Tracks were laid on the frozen ground surface during the winter and ballast trains moved in just before break-up in the spring. These trains operated both ways from ballast pits situated at Miles 396, 467, and 507 and gradually filled the unballasted portions of the line. On October 16, 1928, the line had reached Mile 428 and on April 3, 1929, the last spike was driven at Mile 510 at Churchill. Ballasting followed track laying – the track-lifting gangs raised the tracks with jacks and gravel was pounded under the ties. During the summer extensive ditches were excavated to drain water from the small lakes across which the railway had been built in order to retard thawing of the underlying permafrost and thus prevent serious settlement. The gravel in the ballast pits had to be thawed with steam points. Obtaining an adequate supply of water in winter was a problem because seasonal frost penetrated to the permafrost table and most of the shallow lakes froze to the bottom (Innis 1930). Although problems were encountered because of permafrost, the unusual method of laying tracks before ballast has proven satisfactory. There were stretches where the ballast subsided badly, due to thawing, and additional gravel had to be added.

North to about Gillam in the discontinuous permafrost zone there are no unusual features in maintaining and operating the line (Charles 1959). It is recommended that drainage ditches be re-excavated periodically, say every ten years to prevent accumulation of water against the slopes of embankments. Cuts, especially those excavated through permafrost, should be kept well drained, and berms should be built at the toe of slopes of embankments and up to two feet above high water level at locations where total drainage is not practicable.

North of the Nelson River to Churchill the permafrost is generally continuous and the roadbed consists almost entirely of embankment on the original ground to avoid disturbance of the permafrost. Ditches were excavated only where it was essential to drain lakes and ponds. Maintenance of the track is comparatively light and permafrost has not caused any difficulties. The chief problem is with bridge piles which heave in the winter but do not settle again during the summer. The piles have had to be cut off annually and some may work themselves almost completely out of the ground. (Personal communication, Canadian National Railways).

In 1957 construction was begun on the 31-mile branch line to Thompson. Permafrost was encountered in a number of cuts through the varved clays. Layers of ice within and between the varves melted necessitating very low side slopes to prevent the clay from sliding and obstructing drainage ditches in the bottom of these cuts. Because time was limited to meet a deadline for track to be laid to the new mine site, these cuts were excavated to a depth of only one foot below normal subgrade. As the frozen ground thawed during the following year, clay worked up through the ballast and caused serious track problems. These were not overcome until sections were removed to permit cuts to be deepened to at least three feet below subgrade. These excavations were then backfilled with coarse-grained material, drainage ditches improved, and the track relaid (Charles 1959).

b / Lynn Lake Railway
The other railroad constructed over permafrost in Manitoba is the line branching from the Hudson Bay Railway to Lynn Lake in the northwest part of the province. In the 1930s a line was built from Cranberry Portage north for a distance of about 35 miles to Sherridon, over discontinuous permafrost. At Sherridon a base metal mine operated for about twenty years and then shut down. In August, 1951, a contract was awarded to build 144 miles of railroad northward to the new nickel-copper deposits at Lynn Lake.

The line was built through the Precambrian Shield between latitudes 55° N and 57° N on discontinuous permafrost. Some hills, through which cuts had to be made, had overburden of perennially frozen varved clay and/or peat. Excavation of the frozen material was effected by removing the surface vegetation and scraping the underlying material, layer by layer, as it thawed downward. Where time was limited, explosives were used to excavate some of the deeper cuts. All the cuts in permafrost were excavated to a depth of two to three feet below subgrade to permit the use of more ballast. The sides of the cuts were widened to very gentle slopes similar to the Thompson line (Charles 1959).

Trouble was experienced with the station at Lynn Lake built in 1967. At the site, perennially frozen fine-grained soil with high ice content was excavated to a depth of 18 feet and replaced with sand and gravel. The building was set on a 14-inch concrete slab containing heating ducts. Severe differential settlement has occurred during the two years since construction due to thawing of the ice-laden permafrost remaining beneath the excavation (Personal communication, Canadian National Railways).

The railroad has, since its completion in October, 1953, performed satisfac-torily. Higher construction costs than those normally experienced in non-perma-frost areas were caused mainly by the excavation of frozen soil.

c / Quebec North Shore and Labrador Railway
Construction of this railroad which is owned by a subsidiary company of the Iron Ore Company of Canada began in 1950 and finished in 1954. It extends from Sept Isles, on the north shore of the Gulf of St. Lawrence, 360 miles northward to Schefferville, PQ, in the centre of the Labrador-Ungava peninsula where tremen-dous deposits of high-grade iron ore are located.

The existence of permafrost along the railway is marginal and a few isolated patches have been encountered (Woods *et al.* 1959, Pryer 1966). During excava-tion of a shallow cut at Mile 245, in September, 1953, frozen silt, presumed to be permafrost, was exposed containing numerous ice layers up to 3 inches thick extending to a depth of 6½ feet. To remove this frozen material, it was necessary to drill and blast to a depth of 6½ feet below the subgrade for a distance of 100 feet. Seasonal frost penetration exceeds 10 feet and patches of frozen ground may persist through the summer. In undisturbed peat areas ice lenses have been encountered in excavations for line poles in mid-September. Such occurrences are considered as marginal permafrost.

d / Great Slave Lake Railway
In the early 1960s Canadian National Railways constructed the Great Slave Lake Railway 430 miles in length – extending from Grimshaw in northern Alberta to Hay River on the south shore of Great Slave Lake and including a branch line to serve Pine Point Mines, 53 miles to the east. The railway, which closely parallels the Mackenzie Highway, was constructed primarily to ship zinc ore concentrate from the mine although it also handles southbound wheat and lumber over the first 150 miles, and provides a rail route to Hay River where heavy commodities are trans-shipped to barges for the voyage across Great Slave Lake and down the Mackenzie River (A. V. Johnston 1964).

Permafrost conditions along the railway location are similar to those encoun-tered along the Mackenzie Highway. North of Grimshaw permafrost islands were encountered, mainly in peatlands at Miles 142, 202, 219, 227, 232, 236, and 245 (in highway mileages; railway mileages are almost identical). North of Mile 250 permafrost is more widespread because of the larger and more fre-quently occurring peatlands and peat areas and decreasing mean annual air temperatures. Permafrost occurs north of Mile 250 in more than thirty separate peat areas which vary in size from a few hundred feet to several square miles.

Although the Great Slave Lake Railway extends through the southern fringe of the permafrost region, permafrost was not an important construction consideration (A. V. Johnston 1964). The peatlands which contain most of the permafrost, were avoided as much as possible, and high fills were placed on north-facing river banks where permafrost was encountered. The normal procedure in construction was to leave the existing moss cover undisturbed beneath fills exceeding 4 feet in height. Where permafrost was found in excavations, it was left undisturbed and covered as soon as possible with gravel fill. This procedure was followed at about a dozen culvert sites using gravel pads 3 feet thick.

The marginal economics of branch railway construction forbad any thoughts of excavating small islands of permafrost and backfilling with gravel. Unless there was considerable ice content in the frozen ground it was less expensive to make up any small settlements which occurred by surfacing track during operations. No settlement of grades or heaving of bridges or culverts has so far been attributed to thawing permafrost.

3 Airfields

Northern Canada is served today by a system of air routes the development of which began after the First World War (see Figure 34). In 1926 an airfield was built at Whitehorse, YT, and in the early 1930s strips were built at Watson Lake, YT, and Fort Nelson, BC. In 1943, eight additional airfields, constituting the Northwest Staging Route, were built along the Alaska Highway and operated by the Royal Canadian Air Force. Six of these were east of Whitehorse, in the southern fringe of the permafrost region, and the other two were west of Whitehorse where permafrost is widespread. At the latter two areas large sites having coarse-grained soils were found and no permafrost problems were reported. East of Whitehorse there was no mention of permafrost in the construction reports and it appears that the permafrost either caused no problem or was avoided (Roads and Streets 1944).

During the wartime period and shortly afterwards, airfields were also built at a number of settlements on Great Slave Lake and the Mackenzie River including Hay River, Yellowknife, Fort Resolution, Fort Providence, Fort Simpson and Norman Wells, some of which supported the Canol Project. These settlements are located in the discontinuous zone and permafrost was not encountered at all sites. The airstrips are located mostly on fine-grained frost-susceptible soils which are widespread in the Mackenzie River valley and serious drainage and frost action problems were experienced in their construction and operation.

In the late 1940s an attempt was made to construct a strip at Aklavik in the Mackenzie River delta where adverse soils and permafrost conditions exist. A bulldozer was used to strip the vegetation and thaw soil down to the permafrost table. In a very short time, the perennially frozen, ice-laden silts thawed, turning the stripped area into a quagmire in which the bulldozer was mired. This area was

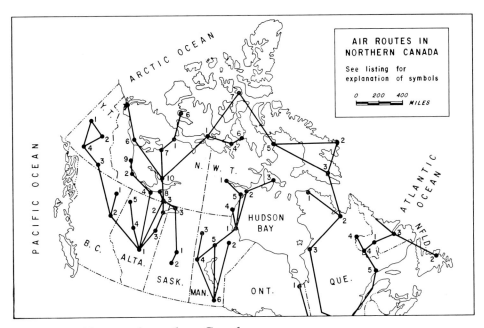

FIGURE 34 Air routes in northern Canada.

Explanation of numbers on map

ALBERTA
1 Edmonton
2 Fort Chipewyan
3 McMurray
4 Peace River
5 Rainbow Lake

BRITISH COLUMBIA
1 Fort Nelson
2 Fort St. John

MANITOBA
1 Churchill
2 Gillam
3 Lynn Lake
4 The Pas
5 Thompson
6 Winnipeg

ONTARIO
1 Moosonee

NEWFOUNDLAND
1 Churchill Falls
2 Gander
3 Goose Bay
4 Wabush

NORTHWEST TERRITORIES
Franklin District
1 Cambridge Bay
2 Cape Dyer
3 Frobisher Bay
4 Gjoa Haven
5 Hall Lake
6 Holman Island
7 Resolute

Keewatin District
1 Baker Lake
2 Chesterfield Inlet
3 Coral Harbour
4 Eskimo Point
5 Rankin Inlet
6 Spence Bay

Mackenzie District
1 Coppermine
2 Fort Simpson
3 Fort Smith
4 Hay River
5 Inuvik
6 Norman Wells
7 Port Radium
8 Resolution
9 Wrigley
10 Yellowknife

QUEBEC
1 Deception Bay
2 Fort Chimo
3 Great Whale River
4 Schefferville
5 Sept Isles

SASKATCHEWAN
1 La Ronge
2 Prince Albert
3 Uranium City

YUKON TERRITORY
1 Dawson
2 Mayo
3 Watson Lake
4 Whitehorse

never used as an airfield and the initial mistake of stripping the surface could only have been rectified by placing a tremendous quantity of coarse-grained soil over the area and allowing the permafrost regime to be re-established. Because there is no such fill material near Aklavik this could have been accomplished only at great expense. As a result Aklavik is isolated during break-up and freeze-up of the Mackenzie River when neither float nor ski-equipped aircraft can land on the river. A small airstrip exists which can be used most of the year by light aircraft on wheels.

At the town of Inuvik, on higher ground adjacent to the east side of the Macken-zie Delta, a special design was used for the airstrip constructed in 1958, to counter-act permafrost conditions. The trees were cut by hand and laid on the ground after which a minimum of eight feet of crushed dolomitic limestone fill from a nearby quarry were placed on the undisturbed ground surface (see Figure 35). Tempera-ture measurements in and beneath the fill ten years after construction reveal that the permafrost has aggraded to the original ground surface and even into the fill. No difficulties attributable to permafrost or any other cause have been en-countered and the airstrip has performed and is in excellent condition.

Airfields have been constructed at the weather stations and other settlements in the Arctic Archipelago in the continuous permafrost zone. The largest airfield in this region is located at Frobisher Bay, NWT. In 1959 modernization was carried out involving lengthening the 6,000-foot runway to 9,000 feet and widening it from 150 feet to 200 feet. During this work permafrost caused problems in a cut 600 feet wide chiefly through bouldery till up to 18 feet deep in some places. Massive ice layers, up to 2 feet thick, ranging from 10 to 300 feet in length were en-countered unexpectedly in the excavation work. To eliminate further trouble the ice was blasted and removed, and the resulting cavities were filled with gravelly material (*Roads and Engineering Construction* 1960).

At Eureka, NWT, a weather station on Ellesmere Island in the northern part of the continuous permafrost zone, an airfield was constructed on perennially frozen silt containing large quantities of ice. Extensive differential thaw settle-ment occurred rapidly causing the surface to become very uneven.

4 Bridges

Bridge maintenance problems caused by thawing permafrost have occurred on the Alaska Highway, particularly in the west part of the Yukon Territory where it crosses the White and Donjek Rivers. The valleys of these rivers are wide with fast-flowing braided streams and gravel bottoms. The driving of piles was greatly hampered by the compacted gravel and by frozen subsoil at the ends of the bridges and beneath the stream beds. To prevent frost heaving, piers were placed at a minimum depth of seventeen feet below the stream bed. Settlement of piers and piles due to thawing of permafrost has occurred at other bridges on the Alaska Highway and other roads. As mentioned previously icings also occur periodically at many bridge sites on the Alaska Highway.

5 Pipelines

During and after the Second World War a number of oil pipelines were constructed in Canada's permafrost region. In the latter years of the war the Canol Pipeline was constructed from Norman Wells on the Mackenzie River southwest across the Mackenzie Mountains to Whitehorse, YT, a distance of 620 miles. This line carried Norman Wells crude oil to a refinery in Whitehorse. A tank farm was built at Norman Wells and another at Whitehorse for storing the oil. A pumping station was built every 50 miles along the Canol line to move the oil. The Canol line met the Alaska Highway at Johnson's Crossing on Teslin River and followed the highway 80 miles to Whitehorse. Gasoline pipelines were built along the highway to Fairbanks, Alaska, a distance of 600 miles and southeast along the highway to Watson Lake, a distance of 300 miles. Another section of pipeline

FIGURE 35 Airfield at Inuvik constructed by placing layers of crushed dolomitic rock fill on the ground surface from which the trees have been cut by hand to minimize disturbance of underlying continuous permafrost.

ran from Whitehorse to Skagway, 110 miles distant. A cutoff from this last line went out from Carcross to connect with the Whitehorse-Watson Lake gasoline line (Richardson 1944a).

The Canol pipeline had to be laid on the ground surface because of the possibility of frost heaving in the active layer (Richardson 1944). Actually this simplified construction although problems arose when the active layer thawed in summer and the pipe settled differentially causing breakage in several places. The oil was subjected to much lower winter temperatures on the ground surface than it would have encountered in the ground. Although the crude oil has a low pour point (about $-70°$ F), considerable pressure was required to maintain flow. A more viscous oil would have had to be heated or housed in a heated utilidor on the surface causing a great deal of extra construction and maintenance expense.

In Norman Wells itself, the only producing oil field in the permafrost region of North America, there is a network of pipelines, laid on the ground surface, to carry oil from the well heads to storage tanks and the refinery. Those carrying crude oil are heated in winter with steam-traced lines but those from the refinery do not have to be so equipped because the oil products are less viscous and warmer. Some line breakages have occurred but these have probably not been caused by permafrost thawing. Gas is used for heating and cooking in some parts of the settlement and is piped in a ¾-inch line laid on the ground surface.

In 1954 a pipeline was built from Haines, YT, to Fairbanks, Alaska, a distance of 626 miles. The terrain included bedrock outcrops, unfrozen gravels and sands, and perennially frozen peat. Frozen ground was encountered in the low-lying areas and occasionally on the tops of hills. The melting of large quantities of ice when the perennially frozen fine-grained soils and peat thawed, due to clearing of the right of way of the pipeline, resulted in quagmire conditions and equipment wallowed in the slurry. Approximately 122 miles of pipeline had to be laid in permafrost terrain. A standard ditching machine had to be modified to dig in the frozen ground and blasting was required in many areas to ditch in bedrock and permafrost (United States 1956).

The rapidly increasing oil exploration activities in recent years in the Mackenzie River and the Arctic Archipelago and the discovery of a major field in Alaska have prompted investigations on the problems of constructing and operating pipelines in permafrost regions. Imperial Oil constructed an experimental four-inch pipeline at Inuvik adjacent to the main access road between the town and the airport (Watmore 1969). Right-of-way clearing in August 1967 began with hand brushing of the trees followed by stripping of the surface by bulldozer down to the permafrost. The trees were piled and many were later put through a chipper for use as surface insulating material. Working conditions were muddy during line laying because of thawing of the permafrost. Surface vegetation removed by stripping was pushed back over the line in irregular piles by bulldozer following installation of the pipe.

Surface conditions were investigated one year later to assess the disturbance to the permafrost caused by installation of the line. Disturbance of the working surface on either side of the line caused ground settlements of 1 to 1½ feet and the underlying permafrost table dropped 2 to 3 feet below its original position. A considerable amount of water collected in these areas from melting of the large quantities of ice in the top layers of the permafrost. Much less degradation of the permafrost had occurred beneath the line itself, particularly sections covered with wood chips. These preliminary observations indicate the large-scale thermal erosion which can result from disturbing permafrost containing large quantities of ground ice, and the effectiveness of insulation such as wood chips, in protecting the permafrost. Future experiments include pumping oil through the line to test its performance and the effects of an operating line on the permafrost.

In anticipation of future production in northern Canada, the federal government is examining the possibility of using pipelines and other methods for moving oil to markets in the south. Oil companies are also working on this problem. When oil was discovered at Prudhoe Bay in northern Alaska in the summer of 1968 considerable interest arose in the possibility of building a pipeline from the field up the Mackenzie River and south to Edmonton to supply the developing deficit area in the American Midwest. However, the decision was announced early in 1969 to build a line across Alaska to Valdez on the south coast and ship the oil by tanker to the west coast of the continental United States. This announcement has apparently postponed development of the Mackenzie River route but it is nevertheless being considered in the light of future probable discoveries in the Canadian Arctic.

C CONCLUSION

There are at present relatively few land transportation routes in Canada's permafrost region and the frozen condition of the ground on which they are located has had considerable influence on their construction and maintenance. As additional routes are located and existing ones extended, especially with the recent accelerated development of the North, permafrost will continue to be a factor which influences the selection of the route and the techniques used in attaining suitable performance.

Once the decision has been made to build a transportation line such as a road or railway, a detailed site selection to determine the best particular location will be required. Except in the southern fringe of the permafrost region, perennially frozen ground cannot be easily avoided. Thus the properties of the soils in the frozen state must be considered, coarse-grained soils being utilized where possible both for a base and/or for fill material. If such soils are not available, there is no choice but to utilize special techniques to counteract the adverse effects of permafrost in fine-grained frost-susceptible soils. Construction will be more costly than on similar soils in temperate regions where permafrost does not exist.

These higher costs will be met, however, if the roads or railways are considered essential as they are in all the routes already built in Canada's permafrost region. The Alaska Highway had to be built from Dawson Creek, BC, to Alaska, and permafrost could not be avoided. The Canol Road and pipeline started at Norman Wells, NWT, where permafrost is widespread, and had to be built southwest to Whitehorse, YT, where permafrost is patchy. The Hudson Bay Railway had to be built to Churchill, Manitoba, over widespread and continuous permafrost.

In the foreseeable future, it does not appear that the density of land transportation routes in the permafrost region of Canada will approach that existing in temperate regions. One reason is the combination of severe climate and the fact that many exploitable resources in the North cannot compete economically with similar ones in more accessible areas. Permafrost is not the only factor affecting northern development nor is it the only one discouraging it. Northern development will proceed in any event, but permafrost will continue to exert a definite influence and the economics of the situation will have to be considered.

7 Mining and Oil Production

Mining is being conducted presently at only a few scattered locations in the permafrost region of Canada and virtually all in the Subarctic. In terms of the total national income derived from this economic activity, the extraction of mineral resources in these remote areas is almost insignificant, mainly because of the severe physical conditions, distance from markets, and high development and operating costs.

However, the mineral resources of the Canadian north are potentially very promising. Ferrous and non-ferrous minerals occur in parts of the Precambrian Shield which stretches from the Atlantic seaboard westward for 2,000 miles to the eastern fringe of the Mackenzie River valley. This lowland, lying on a complex of Palaeozoic strata between the Shield and the Western Cordillera, is believed to be the locale of entrapped pools of petroleum. To the west, in the western portion of Mackenzie District and in Yukon Territory, are the Tertiary folds of the Cordillera containing metalliferous ore bodies and gold-bearing placer gravels. The latter were enriched by post-Tertiary stream action and untouched by Pleistocene glaciation that scoured some portions of the Cordillera and the Precambrian Shield to the east. The Innuitian region in the Queen Elizabeth Islands of the Arctic Archipelago resembles the folded belt of the Appalachians and contains anticlinal structures of Devonian age which hold promise of oil.

Most types of mining require large capital investment and heavy equipment. Some prospecting can be done with a minimum of equipment, and already large areas have been staked, but there is, however, a wide gap between this initial stage and exploratory drilling, and an even wider one before production.

The advent of the airplane in the Canadian North revolutionized mining activity, particularly at the prospecting stage. The heyday of the lone prospector has gone although a few individuals still hope to make the "big strike." Land, water, and air transport is used and required to move in heavy equipment for both drilling and production. Water and land routes into the north are limited in number and the former are open for only a short period in the summer. In some cases a land route has been built solely to exploit the mineral resources of a particular district. In other cases, however, a route may have to be established to encourage or assist mining companies in exploration and production. The extracted minerals have to compete with similar ones produced closer to the markets in more densely populated areas, and higher production and shipping costs, due to difficulties caused by the severe physical conditions, plus the possibility of uncertain markets, may otherwise make the risk of development too great.Those that have developed have done so in response to world demand and are usually economically competitive or in some cases are subsidized.

The mining activities under consideration here are those situated in the permafrost region of Canada (shown in Figure 36). They are few in number and widely scattered. Inaccessibility, difficult transportation routes, and great distances from world markets in some cases greatly influence these ventures and are the main causes for the present small number. The exploitation of mineral deposits in a perennially frozen state is not impossible but the operation is hindered by the below-freezing ground temperatures and frequent occurrences of ice. Permafrost exists at each of the mining areas discussed in this chapter. Therefore, problems associated with permafrost are common to all mining operations in this vast region, varying in degree from one area to another depending on the type of resource being developed and severity of the permafrost conditions.

The first major mining operations in the permafrost region began west of Hudson Bay with the removal of gold from the placer deposits in the Klondike District of Yukon Territory at the end of the nineteenth century. Production in this area ceased in 1966. A few developments took place between the 1890s and the Second World War. At Carmacks in central YukonTerritory, the Yukon Coal Company Limited began mining coal in 1900. Several operations began in the Mackenzie Valley: oil was discovered in 1920 at Norman Wells on the Mackenzie River ninety miles south of the Arctic Circle and a refinery has been operated by Imperial Oil Limited since that time. In 1930 a radium mine was opened at Port Radium, at the eastern end of Great Bear Lake, which closed down in 1960 because the ore was exhausted. Gold was discovered at Yellowknife in 1938 and the Consolidated Mining and Smelting Company began mining operations there. Northwest of Aklavik near the Mackenzie Delta a small coal mine was opened to supply coal to Aklavik. In northern Manitoba, non-ferrous producers opened at Flin Flon and Sherridon. The Hudson Bay Mining and Smelting Company is still active at Flin Flon but the mine at Sherridon closed down in the early 1950s and the company, Sherritt Gordon Mines Limited, moved to Lynn Lake beginning operations there in 1952.

In the years since the Second World War a number of new mines have come into production west of Hudson Bay. Four new mines have opened in Yukon Territory beginning in 1947 with United Keno Hill Mines Limited, the largest silver mine in Canada; a gold mine at Discovery Mines, west of Carmacks, opened

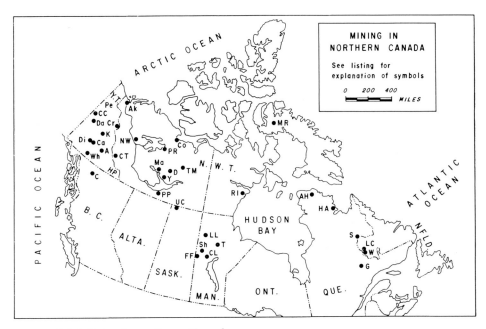

FIGURE 36 Mining in northern Canada.

Explanation of symbols on map

BRITISH COLUMBIA

C   Cassiar

MANITOBA

CL   Chisel Lake
FF   Flin Flon
LL   Lynn Lake
Sh   Sherridon
T   Thompson

NEWFOUNDLAND

LC   Labrador City
W   Wabush

NORTHWEST TERRITORIES

Ak   Aklavik
CT   Canada Tungsten
Co   Coppermine
D   Discovery
Ma   Marian River
MR   Mary River
NW   Norman Wells
PP   Pine Point
PR   Port Radium
RI   Rankin Inlet
TM   Tundra Mines
Y   Yellowknife

QUEBEC

AH   Asbestos Hill
G   Gagnon
HA   Hopes Advance Bay
S   Schefferville

SASKATCHEWAN

UC   Uranium City

YUKON TERRITORY

A   Anvil
Ca   Carmacks
CC   Clinton Creek
Cr   Crest
Da   Dawson
Di   Discovery
HP   Hyland Plateau
K   Keno
Pe   Peel Plateau
Wh   Whitehorse

in 1964; in 1966 New Imperial Mines Limited began producing copper near Whitehorse; and in 1967 the Clinton Creek property of Cassiar Asbestos Corporation sixty miles northwest of Dawson came into production.

Several new mines have been established in the Mackenzie District of the Northwest Territories. In 1948 the largest gold mine in northern Canada, Giant Yellowknife Mines Limited, began production at Yellowknife. This was followed in 1950 by Discovery Mines Limited at Giauque Lake northeast of Yellowknife which terminated production in 1969. Uranium production began at Marian River, at the north end of Great Slave Lake, in 1957 but closed after a few years because of diminishing ore reserves. Another gold producer, Tundra Gold Mines Limited, in the continuous permafrost zone, began producing 200 miles northeast of Yellowknife in 1964. The same year a silver mine began production at the eastern end of Great Bear Lake using the mill at Port Radium which had been closed in 1960. The first tungsten mine in Canada was opened in 1964 in the southwest corner of Mackenzie District near the Yukon Territory boundary but in 1966, a fire which destroyed the mill, interrupted production until 1968. The most recent development in the Mackenzie District is Pine Point Mines operated by Consolidated Mining and Smelting Company of Canada Limited which began production of base metals in 1965 southeast of Hay River, near Great Slave Lake. Also in the Northwest Territories, but near Hudson Bay in Keewatin District, is North Rankin Nickel Mines Limited which began production in 1957 but shut down in 1963.

Mining developments have also taken place in the northern parts of the western provinces in the discontinuous permafrost zone. The first mines to go into production after the Second World War were producing uranium on the north shore of Lake Athabaska where more than a dozen mines were operating soon after discovery of this mineral in 1948. Within a few years most of them closed because of ore depletion leaving only two which operate at present. In 1955 the Cassiar Asbestos Corporation began producing asbestos in northern British Columbia and in addition several large mining developments have taken place in northern Manitoba, e.g., Sherritt Gordon Mines Limited began producing base metals at Lynn Lake in northwestern Manitoba in 1952. In the late 1950s a nickel mine was established at Thompson halfway between Lake Winnipeg and Churchill, the only place in the world where mining, smelting and refining of nickel are all carried out on the same site. About the same time Hudson Bay Mining and Smelting opened new base metal mines at Snow Lake and Chisel Lake about fifty miles east of Flin Flon.

East of Hudson Bay no mining development took place prior to the Second World War. Since the war several large iron mines have come into production in the Labrador Trough which extends in a north northwest–south southeast direction through western Labrador and north central Quebec. The Iron Ore Company of Canada opened the first mine at Schefferville in the early 1950s followed in subsequent years by this and other company developments of new iron mines at the new towns of Wabush and Labrador City in Labrador, and Gagnon in Quebec.

Several very promising areas are presently being explored for mining possibilities including a number in the continuous permafrost zone. The Yukon Territory is the scene of considerable activity where many highly mineralized areas are known to exist. The Anvil Mining Corporation is developing a lead-zinc mine in the Vangorda Creek area in eastern Yukon Territory. Further north, near the Arctic Circle, Crest Exploration Limited has discovered extensive deposits of hematite. Oil exploration has been carried out at two locations in Yukon Territory: in the southeast corner on the Hyland Plateau and in the north on the Peel Plateau.

In the Northwest Territories there are several areas of promise for mineral and oil exploitation. Extensive deposits of copper and other minerals have been discovered in the continuous permafrost zone thirty miles south of Coppermine. In northern Baffin Island in 1962 vast deposits of high-grade magnetite iron ore were discovered. An active exploration programme was carried out from 1963 to about 1965. Oil drilling and exploration have been carried out in Mackenzie District along the Mackenzie River and in the delta, and adjacent areas along the arctic coast. In addition, oil drilling rights have been bought for extensive areas in the Arctic Archipelago and three exploratory holes were drilled prior to 1969.

Exploration is also being carried out in the discontinuous permafrost zone in the northern parts of the provinces, particularly British Columbia and Quebec. The geological framework in British Columbia is similar to Yukon Territory and equally promising mineralized areas are being investigated. In Quebec, exploration in the iron-rich Labrador Trough continues. In addition to this an active search for non-ferrous minerals has produced several promising areas, amongst them a rich deposit of asbestos found in the northwest corner of Quebec in the late 1950s where production is expected to begin in the 1970s.

Mining exploration and production activities in the permafrost region of Canada are increasing rapidly. Before the Second World War there were only about six mining areas in northern Canada, since then about twenty new mines have been brought into production and many rich deposits discovered. Though the permafrost problems encountered by mines in operation will be faced by many of the new producers, some have not found permafrost to be a problem or have not reported their difficulties. This chapter reviews the difficulties which permafrost, as a potential problem in all mining operations in northern Canada, has caused.

A DEVELOPMENTS BEFORE THE SECOND WORLD WAR

1 *Yukon Territory*

a / Placer gold

The first mining operation in Canada's permafrost region was carried out in the Klondike District near Dawson, YT, close to the northern boundary of the discontinuous permafrost zone. The presence of perennially frozen ground in the placer gold deposits, derived from river gravels, caused difficulties from the early stages of development.

The main production of placer gold in Canada has been from the Klondike, but prospecting was conducted in Yukon Territory for fifteen years or more prior to the big discovery in 1896. Discovery of this very rich field was delayed because attention was directed initially to larger streams like Yukon, Stewart, and Big Salmon rather than to smaller streams from which most of the Klondike gold has come (Cooke and Johnston 1933).

Some placer gold has also been produced from the Fortymile, Sixtymile, Mayo, Big Salmon, and Kluane districts and other isolated creeks, as well as the Klondike. Because of the unprecedented rush to the Klondike in 1896 and succeeding years, more attention has been focused here than on any of these other areas. The following discussion of the role of permafrost in Yukon placer gold mining will be limited to the Klondike District, an area of approximately 800 square miles in the northwest corner of the Klondike Plateau section of the Yukon Plateau in the Western Cordilleras. It is bounded by Yukon River on the west, its tributaries, the Klondike and Indian Rivers, on the north and south respectively, and tributaries of these latter two rivers on the east. The plateau is dissected by streams and consists of sloping ridges radiating from a point near the Dome, the highest point of the district. This dome, which is the main drainage centre of the district, is nineteen miles southeast of Dawson about midway between Indian and Klondike Rivers. Its summit is 4,250 feet above sea level, 3,050 feet above Yukon River at Dawson and approximately 500 feet above the ridges at its base. Several gold-bearing creek valleys radiate from it (Cooke and Johnston 1933; Bostock 1948).

Geologically the plateau consists mainly of Precambrian schists pierced by igneous intrusions. Overlying these schists are Tertiary sedimentaries and volcanics and subsequent superficial accumulations of unconsolidated material. These include thick deposits of silt in the stream valleys on top of which may be found varying thickness of gravels. Stream action has formed terraces in these materials on the valley slopes.

Klondike District was not glaciated during the Pleistocene. The gold-bearing gravels were not covered with glacial drift nor disturbed or eroded by an overriding ice sheet. Pleistocene glaciers in nearby areas extended down Lewes River only to Rinks Rapids, leaving the Klondike and surrounding regions ice free. A thick mantle of deeply weathered rock, usually intermingled with talus, covers the valley slopes nearly everywhere (Cooke and Johnston 1933, p. 58).

The cross-section of a typical gold-bearing stream valley in the Klondike shows usually a comparatively narrow inner valley, bordered on one or both sides by wide benches above which the surface rises gradually to the crests of the divides. These benches are remnants of old valley bottoms partially eroded during the formation of the present valleys. Narrow terraces occur in places between the level of these old channels and the level of the present stream. Gold-bearing gravels are spread on the present valley bottoms, on the rock benches cut into the valley sides, and in the preserved areas of the old high level benches (Cooke and Johnston 1933).

The low-level Creek gravels, of Quaternary age, are the most important gold-bearing ones. Found on valley bottoms to depths of 4 to 10 feet, they rest unconformably on the bedrock and in many places the gold extends down cracks and joints in the rock to depths of 2 to 3 feet or more. Therefore in mining a few feet of bedrock must be removed. This zone consists of decomposed and broken schists overlaid by frozen "black muck" – a local term used to refer to organic silt, i.e., silty soil containing large quantities of intermixed partially decomposed vegetation – 2 to 30 feet thick, containing large quantities of ice. In 1938, all the dredging done by the Yukon Consolidated Gold Corporation, the largest operator in the Klondike District, was in the Creek gravels.

The low-level River gravels, containing gold in paying quantities, were on the wide flats bordering the lower reaches of Klondike River below the mouth of Hunker Creek. These gravels were enriched by the deposition of gold transported by tributary streams and derived from former extensions of the high level White Channel gravels, which were eroded away by the deepening of the Klondike River valley.

Three main conditions account for the richness of the gold placers in the Klondike District. First, the bedrock is mineralized; it contains numerous quartz veins carrying small amounts of gold. Second, the region is a deeply eroded plateau remnant; probably several thousand feet of rock have been removed by stream action and the contained gold was concentrated in gravels formed from more resistant rocks such as vein quartz. Third, the Klondike was not glaciated during the Pleistocene. Therefore, gold concentrated in the various series of stream gravels remained there and was not swept away or scattered by glacial ice, or buried under glacial debris (Cooke and Johnston 1933).

Permafrost varies in thickness from a few feet in some localities to more than 200 feet. This maximum thickness encompasses the surficial muck deposits, gold-bearing gravels, and the underlying bedrock on which the gravels lie. The thickness of the perennially frozen ground is less on ridge tops than in the valleys and less on south-facing than on north-facing slopes because of the greater quantity of solar radiation received by the former.

Subpermafrost water was encountered in valley bottoms and this caused difficulty in sinking deep shafts. The sinking of a shaft on a ridge south of Eldorado Creek was stopped by running water at a depth just exceeding 200 feet. Another shaft sunk through gravel on the upland plateau between Bonanza Creek and Klondike River passed into unfrozen ground 75 feet below the ground surface. Near the head of Quartz Creek, running water was encountered in a shaft at a depth of 216 feet (Tyrrell 1903).

Most of the gold in the Klondike District was in gravels sandwiched between the muck and the bedrock. Some gold lay in the top weathered portion of the underlying bedrock where it was deposited in surface cracks. Because permafrost extended from near the top of the muck into the gravel and often into the bedrock, the main problem of exploitation was the removal of the frozen muck and frozen gravel. Through the years, from the first mining at the end of the nineteenth century, a variety of methods have been employed to thaw and remove the frozen material.

The first prospectors relied on solar radiation to thaw the ground. The layer of ground above the permafrost, thawed by the sun, was removed, next day another few inches had thawed and was removed, and the process was repeated daily until a shaft had been sunk to the gold-bearing material. This laborious method was used before the Klondike discovery and produced low yields because mining could be conducted only during the short summer season (Berton 1956).

The abundance of forests, particularly white and black spruce along the main streams, prompted the use of wood fires to hasten the thawing process. Instead of wintering in town, the miner remained on his claim cutting wood with which he built fires to thaw the frozen ground. It was sluiced as soon as stream water was available in the spring. In this way, the work output was more than doubled. The main drawback was the limited supply of wood, except on the main rivers (Innis 1936).

In the winter of 1898–9, it was discovered that frozen gravel could be thawed at the rate of eight to ten feet per day with steam. Steam was forced into the frozen ground with iron pipes connected to a steam boiler. The miner drove the steam points into the ground where they were left for a short time until the ground thawed. The thawed muck, containing a large amount of melted ice, turned a steamed area into a quagmire. Berton reported the scene as: "Seas of hot bubbling mud, jets of steam rising from the ground all over the valley, men standing on step-ladders swinging their sledges, pressure building up as the ground baked around the base of the point. Occasionally it exploded and a point driver was scalded and sometimes killed" (Berton 1956, p. 132).

The main change in the landscape resulting from the intensive mining activity was the rapid depletion of the timber resources of the Klondike District and the denudation of many hills and valley bottoms. Wood was the only fuel available for fires and steam-boilers and by 1909, after a decade of mining, little timber remained on the principal creeks of the district (Canada 1909).

The next major change in thawing technique occurred in 1917 and 1918 when it was discovered that ground sluicing or hydraulicking using cold water would thaw frozen ground. As a result, many low grade placer deposits, formerly ignored because of the high cost of steam thawing, became economically important with the development of these cheaper cold water thawing methods (Lund 1951).

Therefore, since the first mining of frozen gold-bearing gravel was begun in the latter part of the nineteenth century, there has been continual evolution in the methods used to thaw the frozen material. Natural thawing by solar radiation was superseded by artificial thawing methods, first, wood fires on the ground surface followed by steam and hot water, and, finally, cold water. Actually, all these methods except wood fires were used until mining ceased in 1966, depending on the physical properties of the gravel and the schedule of operations. For example, the slowest method, natural thawing by solar radiation, was used where deposits were to be mined two or three years after initial thawing began, because it was the cheapest method. Artificial thawing, which was costlier but quicker, was used where deposits were to be mined the same year. Consideration was also being given to thawing by flooding several years (five or so) in advance of dredging.

After the frozen ground was prepared by these various methods of thawing, it was removed manually by individual miners or, where mining operations were on a large scale, by dredges. The first dredge in Yukon Territory was a small (3¼ cubic feet) bucket dredge built on Lewes River in 1899. In 1901 it was moved and rebuilt on Bonanza Creek and then in 1903 was moved to the Discovery group of claims. This dredge became obsolete in 1905 but it proved the practicability of dredging on the Klondike creeks.

After the richest gold-bearing gravels had been removed by the most profitable and easy method – individual miners – only large companies with sufficient capital to invest in heavy mechanical equipment were able to operate at a profit. By 1930 most of the gold was taken by dredging. By 1950 the Yukon Consolidated Gold Corporation Limited was the largest company in the Klondike District and accounted for two-thirds of the total annual yield of $2,100,100 to $3,600,000 (Collins 1955).

The latest mining procedure used until termination of production in 1966 consisted of three phases described by Patty, 1945. The first was prospecting, the evaluation of the quantity of gold in a given deposit. This operation began about 15 April each year.

The second phase, stripping the ground surface of moss, brush, and timber, was carried out between 15 May and 15 July because the ground became too soft later in the summer for tractor work. In addition, any muck thawed by solar radiation was removed. To keep these operations clear of the thawing and dredging, and to allow the greatest efficiency in mining, stripping was completed about two years ahead of thawing, and thawing a year and a half ahead of dredging.

Where nearby creeks were available to supply a reliable source of water, stripping was done by hydraulicking. Water under pressure was carried through a system of pipes. To obtain the best efficiency from the water, a fairly large area was covered in a single operation to allow solar radiation to thaw the surface of the ground. Hydraulicking was done generally late in the summer when natural thaw had penetrated a foot or more beneath the ground surface.

The third stage, thawing, included both natural thawing by solar radiation and artificial thawing with cold water. Natural thawing was alternated with stripping. After the naturally thawed ground surface material had been stripped, either by tractor or hydraulicking, the exposed frozen muck thawed about 4 inches in the next 24 hours. An additional 2 to 3 inches thawed the second day with progressively less penetration on succeeding days. The key to successful muck stripping was to strip the thawed material by hydraulicking thus continually exposing frozen material to the sun. Because of the high ice content of the muck, the stripping area soon assumed a "badland topography" with many potholes, miniature canyons, and waterfalls. Muck stripping proceeded twenty-four hours a day. Exposed frozen gravels 10 to 20 feet thick usually thawed in three years and were ready for dredging. In this period 2 to 3 feet of the underlying bedrock would also be thawed. On shallow gravels where the material was coarse, two years of exposure were usually sufficient. On deeper gravels, or deposits containing lenses of finer material, longer exposure was required.

To accelerate thawing, cold water was injected into the frozen ground through pipes. Each pipe or "thawing point" was forced down to the permafrost table by hand and then the operator moved to the next point about 15 feet away and repeated this act. A few hours after making the rounds of about fifty points he returned to the first point, by which time the water spurting from the end of the point had thawed the gravel down 3 to 4 feet ahead of it. Again the point was pushed down until the frozen gravel was encountered and the operator moved to the next point.

Each thaw point was advanced several feet per day. Gravel which was 20 feet thick was thawed to bedrock in about twelve days by this method. This operation became hazardous for the operators because water bubbled up at many points on the ground surface and the point driver had to lay poles on the ground so that he could move around from one point to another without sinking into the water-soaked gravel. The thawing water was about 38° F when it entered the ground and 34° F when it returned to the surface.

The thawing season lasted from about 10 May to 25 September, allowing about 120 days of thawing for a full dredging season. By July the air temperature was usually above 50° F and the rate of thawing accelerated.

Dredging followed the completion of thawing. This was accomplished easily when the ground was well thawed, but this state was not always reached and the dredges often had to dig some frozen ground.

A comparison of the cost of various operations in thawed and frozen ground furnishes proof of the additional effort required to overcome the handicaps imposed by permafrost. In 1909 it was reported that in the prospecting stage of the "ground preparation," the cost of drilling was two dollars per foot in unfrozen ground and four dollars per foot in frozen ground (Canada 1909). In 1916, the cost of drifting from a vertical shaft was two dollars per running foot in unfrozen ground and three dollars per running foot in frozen ground. With the discovery of cold-water thawing in 1917 and 1918, thawing costs were halved from those when steam was used. Nevertheless, in 1939 stripping and thawing still were costing from 33 to 50 per cent of the total mining operation. In 1945 it was reported that ground preparation costs alone were costing more than the entire mining operation of similar but unfrozen deposits in California (Patty 1945).

It is interesting to compare conditions in Yukon Territory with the situation in Siberia where large-scale placer gold-mining operations in permafrost are being carried out at present. One thawing method currently employed by Russian operators, which was not used in the Klondike, is the flooding of areas to be mined. These areas are first stripped of muck down to the gravel and then flooded. Water is impounded to a depth of about ten feet behind earth dams arranged in a series of ponds up a stream valley which contains perennially frozen gold-bearing gravels. This procedure is initiated several years before mining is to begin. The warming effect of the water causes the permafrost to thaw down to the depth of operation of the dredges. After thawing, the dredges work one pond at a time up the valley. The Russians report that this method is the most economical for handling the frozen ground.

b / Coal

One coal mine has been producing in Yukon Territory since 1900. The Yukon Coal Company Limited operates the Tantalus Butte Mine at Carmacks in the central Yukon, which supplies bituminous coal to United Keno Mines Limited. The annual output is about 8,200 tons (Dubnie and Buck 1965). Carmacks is located in the discontinuous zone and permafrost is widespread in the mine. Thin layers of ice are known to exist in the coal seams but there is no report of any adverse effects caused by permafrost in the mining operation.

2 Northwest Territories

a / Oil

The only producing oil field in Canada's permafrost region is at Norman Wells, NWT, on the east bank of the Mackenzie River, ninety miles south of the Arctic Circle, in the discontinuous permafrost zone. Oil seepages along the banks of the river had been recorded by Alexander Mackenzie in 1789. The discovery at Norman Wells in 1920 was an important factor in the stimulation of prospecting both for oil in the Mackenzie Valley and for metalliferous deposits in the Precambrian Shield to the east (Dawson 1947).

The oil at Norman Wells is in a Devonian coral reef which is sufficiently porous to form a fair reservoir and the shales that surround it are impervious enough to retain the oil in the reef rock. Wells drilled in this field outlined an area of over 5,000 acres that could be productive. A large part of the field is under Mackenzie River between the north bank and Bear Island about a mile and a half distant.

A small refinery was built in 1932 at Norman Wells to supply diesel oil, gasoline, and fuel oil mostly for the development of the pitchblende deposits at Port Radium. In 1941 the sale of products from Norman Wells was 80,000 gallons of aviation gas, 112,000 gallons of motor fuel, and 230,000 gallons of fuel oil marketed at settlements in the Mackenzie Valley. This production was not the result of war demands because the Canol Project, begun in 1942 and completed in 1944, saw further exploration and drilling in this known oil field. In 1944, when the demand for oil for the Pacific war theatre was at its height, there were 60 producing wells of a total of 67 drilled. More than 1,000,000 barrels of crude oil were produced, supplying gasoline and diesel oil used on the Northwest Staging Route and on the Alaska Highway. After 1945, when war demands ceased, production was reduced considerably to meet the limited demands of settlements in the Mackenzie Valley. After the war, construction of the Distant Early Warning Line and increase in size of such settlements as Yellowknife and Inuvik increased demand and production. In the early 1960s production was nearly 900,000 barrels of crude oil decreasing to about 500,000 in 1964. It rose to nearly 700,000 barrels in 1966 and 862,878 barrels were produced in 1968 (Dubnie and Buck 1965; Personal communication, Department of Indian Affairs and Northern Development).

At Norman Wells permafrost is about 200 feet thick, becoming thinner towards the river. Holes down to the oil vary in length from 1,300 feet for vertical holes to 2,400 for directional holes which are slanted to tap oil under Mackenzie River. The oil lies sufficiently deep below the permafrost that its properties *in situ* are not affected by the low rock temperature overlying it.

Some difficulties have been encountered in obtaining good cementing at depths where the ground temperatures are below 32° F. Generally this difficulty does not arise in deeper holes in which drilling mud has been circulated for some time with subsequent warming of the ground around the hole. Failing this, some method of heating the cement is required to allow it time to set properly. This low temperature at depth also accelerates wax deposition within the tubing well-heads of the flow lines. In such cases, the wax must be removed by heating or by mechanical means such as scrapers (Hemstock 1952). On the surface the crude oil is stored in heated tanks so that it can be pumped to the refinery in winter.

b / Pitchblende

A pitchblende-silver orebody was discovered in 1930 in the Precambrian Shield on the east shore of Great Bear Lake. From 1934–9, the Eldorado Mine produced over seven million dollars of pitchblende for radium, a disintegration product of uranium. The mine was idle in 1941 and 1942 but was expropriated by a crown company in 1944, and the uranium content of the pitchblende became the important mineral. Until 1960, when the mine was closed, the ore was concentrated in the mill on the mine property and then shipped to Port Hope, Ontario, for refining (Dawson 1947).

The mine was situated in the northern part of the discontinuous permafrost zone near the boundary of the continuous zone. Underground workings extended to a depth of 1,350 feet and permafrost was encountered to a depth of 345 feet below the ground surface (Lord 1941). No information on details of mining in the permafrost are available.

c / Gold

In 1934 gold-bearing rocks were discovered in a severely faulted and intruded region of the Precambrian Shield north of Great Slave Lake. The Con Mine on the west shore of Yellowknife Bay began producing in 1938, and the neighbouring Rycon and Negus Mines followed in 1939. Three more mines were producing by 1941 and active development was progressing on others. Work ceased on all of the non-producing properties during the first years of the Second World War and was not resumed on most of them. Negus closed in October, 1944 but resumed milling in 1945. No reports of permafrost are available from these operations although they are located in the northern part of the discontinuous zone.

d / Coal

Coal seams, many of which are low grade, occur at various locations in the

Northwest Territories. Eskimos have mined surface exposures near Pond Inlet in northern Baffin Island.

Low-grade bituminous coal was found prior to the Second World War in the 1930s in the Richardson Mountains about sixty miles northwest of Aklavik, in the continuous permafrost zone. The coal was extracted from a horizontal shaft extended into the side of a hill and the total annual production of about 1,000 tons was marketed in Aklavik to the church missions. Information concerning difficulties imposed by permafrost is scanty but it is known that warm summer air penetrating into the mine caused some thawing and water accumulation on the floor of the shaft. No mining was carried out during the winter. The mine ceased operation about 1960.

e / Base metals
Prior to the Second World War mining of base metals in the permafrost region was underway at Flin Flon and Sherridon – both located in Manitoba near the southern limit of the discontinous zone. The Hudson Bay Mining and Smelting Company Limited is still active at Flin Flon but Sherritt Gordon Company shut down its mine at Sherridon because of ore depletion and opened a new one at Lynn Lake. Flin Flon and Sherridon are both located in the southern fringe of the discontinuous zone where permafrost occurs mostly in peatlands and other special types of terrain. No permafrost has been encountered in the bedrock so it is not a factor in these mining operations.

B   DEVELOPMENTS AFTER SECOND WORLD WAR

1 *Yukon Territory*

a / Base metals
The first mine to begin production in the permafrost region of Canada after the Second World War was United Keno Hill Mines Limited in the central Yukon Territory in 1947. Silver-lead-zinc deposits occurring in commercial quantities support the three settlements of Mayo, Elsa, and Keno Hill located about 220 miles northeast of Whitehorse. Mining first began in the area in the early years of the twentieth century and today United Keno stands as the largest primary producer of silver in Canada (Boyle 1956).

The relief is mountainous with elevations varying from 2,300 to 6,750 feet above sea level. Below 4,500 feet, there are few rock outcrops and the slopes are covered with till, rock debris, and peat on which grow conifers, birch, and aspen. Above 4,500 feet the terrain is relatively flat and rolling with several prominent rocky knolls containing the orebodies. The most important of these, Keno and Sourdough Hills, rise to over 6,000 feet above sea level. Above 4,500 feet, outcrops are numerous, soil is scarce, and the vegetation is limited to alpine varieties (Boyle 1956, Canadian Mining and Metallurgy Bulletin 1961).

The lower slopes of Keno and Sourdough Hills were severely glaciated during the Pleistocene. Till and gravel are widespread and generally 5 to 20 feet thick, but on some places, like the south-facing slopes of Keno Hill, the deposits attain a thickness of 30 to 50 feet. Some valleys are U-shaped and floored with sand, gravel, and till through which streams have cut channels bordered by a series of benches (Boyle 1956; Canadian Mining and Metallurgy Bulletin 1961).

This mining district is located in the discontinuous zone where permafrost is widespread. The mean annual air temperature at Mayo is 26° F but at higher elevations it is several degrees lower causing the extent and thickness of the permafrost to vary considerably. Its distribution depends on elevation, exposure, depth of unconsolidated overburden, vegetation, and the presence of flowing surface and underground water. The maximum thickness of permafrost encountered in mining exceeds 400 feet at elevations of 5,000 to 6,000 feet above sea level.

The ground on the north-facing slopes of Keno and Sourdough Hills is mostly perennially frozen in contrast to the south-facing slopes which are relatively free of permafrost, especially at lower elevations. This is because of the greater solar radiation received by the south-facing slopes. On Keno Hill the mine workings on the top of the hill and on the north-facing slope were in permafrost to a depth of 400 feet below the surface. On Sourdough Hill, ice lenses were encountered in the Bellekeno mine workings 250 feet below the surface. On the lower south-facing slope of Keno Hill, however, the workings of the Mount Keno mine were not in permafrost. The main shaft of Lucky Queen mine enters the ground at 5,100 feet above sea level and extends 360 feet below the surface without passing out of perennially frozen ground. Ground temperatures taken at the 100-, 200-, and 300-foot depths were respectively 28° F, and 29.3° F, and 30.7° F indicating a gradient of 1.3° F per 100 feet (Wernecke 1932).

In the permafrost zones of many mines ice veins have been observed up to 6 inches in width. They occupy fractures in the rock caused by the crystallizing force of the groundwater which froze when the permafrost was being formed. Often they occur in groups up to 5 or 10 feet in width. Although none of these veins has been exposed for its full length, they are short compared to the metalliferous veins. In one stoping operation, an ice vein was encountered that occurred along a fault rising from the 200 foot level to the 50 foot level in a short distance. Those that fill angular cracks are believed to be much shorter with a maximum length of 25 to 30 feet. The ice veins are larger and more frequent between the 100 foot level and the surface than below this level. At the 300 foot level the ice veins are seldom more than ¼ inch wide and inclusions of ice exceeding 2 inches in diameter are rare (Wernecke 1932).

The existence of permafrost causes, at several stages of the mining operation, a number of problems which increase costs. Drilling for exploratory purposes and blasting are hampered by the possibility of drill rod freezing in the hole. Drilling is frequently carried out on a twenty-four-hour basis until the particular hole is completed; calcium chloride is used in the wash water to depress its freezing point. In many instances drill holes with water in them have to be blasted or cleared with compressed air on the same shift or the water will freeze in the holes. When the humidity is high, as in the summer, any moisture in the compressed air tends to condense when it reaches the lower temperatures of the underground workings and results in frozen air lines and machines (Pike 1966).

Some surface mining is carried out. Solar radiation is used for natural thawing as the top few feet of overburden are scraped off each day with a bulldozer (Personal communication, United Keno Hill Mines Limited 1959). Because of the ice content, the rock has a higher elasticity than rock at temperatures above 32° F. As a result, more blasting holes and powder are required. The ore must be blasted and removed in small quantities because large quantities of broken ore tend to freeze into a large unmanageable mass before it can be taken to the surface.

The presence of permafrost causes some difficulties in transporting the ore in the underground workings. Any water coming into contact with the metal of the machines or the tracks freezes rapidly, forming an ice covering which must be removed at frequent intervals. Sand is used to obtain better traction for the locomotives. In summer, warm air circulating through the mine workings melts the ice veins and the walls of the fractures collapse (Wernecke 1932). Because of the danger of this weakened rock falling into the mine drifts and partially blocking them, timber supports are required to strengthen the walls and roofs of the workings. Sometimes it is necessary to re-timber several times because the timbers may move after the warm summer season.

Intrapermafrost groundwater under hydrostatic pressure has been encountered in many parts of the Keno mines. In one shaft, groundwater flooded the lower level until adequate pumping facilities were provided. Eventually a drainage tunnel was excavated which carried away all the underground water in excess of that needed for milling and domestic purposes. In another mine, 100 gallons per minute flowed from a tunnel when the perennially frozen zone was first penetrated. In many cases, springs originating above the mines on the upper slopes of the hills and flowing throughout the year penetrate along faults and vein systems that extend through the frozen zone. Because of this water, drifts which are not worked for a period of time become filled with ice (Wernecke 1932). Sealing aquifers with cement is a problem because of the difficulty of curing cement at the sub-freezing temperatures. It is necessary to heat the rock for a considerable period and then place heated concrete to obtain the required set before the warmed rock becomes frozen again.

b / Gold
Gold mining began at Discovery Mines Limited about 25 miles west of Carmacks, YT in 1964. The underground workings are on a south-facing slope with adits at 3,300, 3,400 and 3,475 feet above sea level. The adits are 2,500 feet long and no permafrost has been encountered in the wall rock. The temperature in the adits is very close to 32° F. Permafrost has been encountered on north-facing slopes and in shaded gullies in road construction. Perennially frozen ground will undoubtedly be encountered in underground workings on north-facing slopes (Personal communication, Discovery Mines Limited 1964).

c / Copper
New Imperial Mines Limited, the first copper producer in Yukon Territory, began mining near Whitehorse in 1966. It is a strip mining operation located in the southern fringe of the discontinuous permafrost zone. No report of permafrost has been received from this mining operation and it is unlikely that it will be a factor since any occurrences of permafrost in the vicinity are found mostly in peat bogs.

d / Asbestos
Cassiar Asbestos Corporation Limited began production of asbestos in October 1967 at its Clinton Creek property, sixty-five miles west of Dawson in the discontinuous permafrost zone, and conditions are similar to those at Dawson where permafrost is widespread and about 200 feet thick. There is no report available on the occurrences of permafrost in the open pit mining operation but it has been encountered at the plant site and the townsite five miles away (Mining in Canada, 1967a, C. J. Brown 1965).

2 Northwest Territories

a / Gold
In 1944 gold-bearing orebodies larger than any previously known to exist at Yellowknife were found on claims of the Giant Yellowknife Mines Ltd. (Dawson 1947). Precambrian bedrock outcrops over about 30 per cent of the Giant property. The remainder is covered by Pleistocene and Recent deposits which vary in thickness up to 110 feet. This unconsolidated overburden consists of several feet of Pleistocene sand and gravel lying on the bedrock overlaid by Recent thinly stratified lacustrine clays. These clays were deposited probably

when Great Slave Lake was at a higher level than now in late glacial time (Bateman 1949).

Yellowknife is located in the discontinuous zone where permafrost is widespread and extends to depths exceeding 200 feet. The mean annual air temperature is 22° F. Most of the present town, except for the first buildings on a rocky knoll at the edge of Great Slave Lake, is laid out on a flat glacial sand and gravel outwash plain. Excavations for sewer and water mains encountered permafrost in clay, but rarely in sand and gravel. Several wells were sunk beneath buildings without encountering frozen ground (Bateman 1949).

At the Giant Yellowknife property on the west side of Yellowknife Bay permafrost was encountered during such surface construction as road building and in exploratory diamond drilling and underground operations. Roads were built by dumping gravel or mine waste on the moss cover to retain the stability of the underlying permafrost, and pile foundations were used for buildings. The piles were greased and wrapped in tar-paper collars to reduce the effects of frost heaving, and then placed in holes steamed to depths varying from fifteen to twenty feet.

Where bedrock outcrops exist permafrost was not encountered either at the rock surface or at depth. Shallow deposits of clay, sand, and gravel on uplands and ridges are not perennially frozen where there is no moss cover. There is permafrost in unconsolidated deposits exceeding six to eight feet in thickness particularly where there is a surface covering of peat and moss. In all known cases, this extends throughout the entire thickness of the overburden to bedrock.

Generally the thicker the overburden, the thicker the permafrost and the more severe the problems associated with this phenomenon. At Number One Shaft where the thickness of overburden above the mine workings does not exceed 40 feet, no permafrost was encountered at the shallowest working level of 200 feet. At Number Two Shaft, one mile north of Number One, permafrost extends throughout most of the 115 foot level and in one part of the mine, where the overburden exceeds 60 feet, permafrost was encountered 280 feet below the surface. There are places where the overburden exceeds 100 feet in thickness and permafrost may exceed 280 feet in thickness. It appears, therefore, that the overburden is an insulating blanket preserving the permafrost that may have formed during late glacial time in climate colder than today. Permafrost must have extended into the bedrock to depths at least greater than it is presently known before the blanket of unconsolidated overburden was deposited (Bateman 1949).

Several permafrost problems occur at Giant Yellowknife and various remedial measures are employed. One cut and fill stope is overlain by overburden of gravel, sand, and clay with a ground surface cover of peat and moss. This overburden averages about 40 feet in thickness and permafrost was encountered to a depth of 250 feet. Considerable high grade gold ore was discovered in a horizontal surface pillar of this stope area at the rock-overburden interface. During the extraction of this ore the perennially frozen overburden had to be maintained in a frozen condition to prevent its collapse. Before the mining of this pillar was begun, all unnecessary openings into the stope were sealed with reinforced concrete bulkheads. The extraction of this ore was done during the winter when air in the stope could be more easily maintained at sub-freezing temperatures. Because the overburden was frozen, not many shoring timbers were required to support this material long enough to remove the ore. In view of possible eventual thawing, the remaining openings to this pillar were sealed off when the ore was removed in order to isolate the stope completely from the other mine workings (McDonald 1953).

Permafrost has alleviated one problem, however, in providing conditions for the safe storage of arsenic trioxide, a by-product of the roasters. This material has accumulated at an increasing rate: 10 to 12 tons daily in the early 1950s to almost 25 tons daily since 1958 – reading a total of approximately 140,000 tons by 1969, which occupies a volume in excess of four million cubic feet. It has to be stored where the risk of contaminating surface and underground workings is minimized. Storage space is provided in special underground chambers which were excavated in bedrock some distance from the hanging wall of the ore zones. The chambers are accessible only from the surface, and have been completely sealed off from the other underground workings by massive reinforced concrete bulkheads. Here the overburden exceeds 50 feet in thickness and permafrost extends 270 feet below the ground surface. In this perennially frozen zone, ground water movement, which could carry the arsenic trioxide into mine workings is prevented by the subfreezing temperatures and there is no contamination (Espley 1969, McDonald 1953).

Gold was also mined until 1969 at Discovery Mine 60 miles north of Yellowknife where the permafrost is 250 feet thick. On one occasion the mine had to be closed down for a lengthy period when it flooded and all the openings became filled with ice. It was reopened by siphoning water from a pond into the mine, melting the ice, and pumping the water out again. During the past 22 years of continuous operation the permafrost has retreated 5 to 10 feet from the walls of the excavations (Kilgour 1969).

Gold mining was commenced in 1964 by Tundra Gold Mines Limited, 150 miles northeast of Yellowknife in the tundra north of the treeline. This mine is located in the continuous zone where permafrost has been observed to a depth of 900 feet. The mean annual air temperature is about 20° F and the ground temperatures vary from about 27° F on the first level at the 175-foot depth to 32° F on the sixth level at 925 feet.

Below the permafrost zone there is a steady water flow of about twenty gallons per minute. The permafrost sections of the mine are dry except for summer run-off. Problems occurred when a ventilation raise through to the surface filled with water although the foot of the raise and connecting drifts were dry. The water in the raise was caused by an ice bridge of unknown origin forming about twenty feet down from the collar.

Handling run-off water in the permafrost sections of the mine requires special techniques, especially in long drifts of 3,000 feet. Hot water from the power plant circulating system is pumped underground. Bleeders are inserted in these pipelines at intervals and allow sufficient hot water to mix with the ditch water to keep the ditch open and free of ice. Aluminium pipe is used because sufficient heat radiates from it to raise the temperature in the drifts and cross-cuts in the permafrost zone (Boulding 1963).

b / Base metals

Mining of nickel and copper by North Rankin Nickel Mines Limited began in 1957 at Rankin Inlet on the west coast of Hudson Bay at tidewater, 320 miles north of Churchill (Hannah 1961). Production ceased in 1962 when the high-grade deposits were mined out and the mine closed in 1963 because an exploration programme failed to turn up further ore reserves. The mine is located in the continuous zone where permafrost is about 900 feet thick. The mean annual air temperature is about 17° F and the temperature in the mine is about 27° F. The ore was a copper-nickel sulphide body containing metals of the platinum group (Weber and Teal 1959).

The existence of permafrost affected the mining operation in a number of ways. For drilling water it was necessary to use a heated salt solution which tended to corrode the drills and caused higher maintenance cost – the machinery had to be brought to the surface frequently for lubrication and overhauling. Special atten-tion was also given to the compressed air lines to prevent their being clogged with hoar frost. This was prevented by passing vapourized anti-freeze through the lines (Weber and Teal 1959).

The 27° F temperature of the air underground necessitated that all air taken into the mine was heated. This was accomplished by steam coils in the shaft house: the air was transferred by ducts to the first level and then through the workings to the second level and the passage of this warm air through the workings resulted in a retreat of the permafrost in the walls and a gradual rise in the underground temperature.

Underground transportation was one of the most seriously affected phases of the mining operation. The use of diesels required large volumes of fresh air in the headings and this entailed the heating of large volumes of air through a wide temperature range, with high costs resulting.

Although seepage in the mine was small compared with mines such as Giant Yellowknife and United Keno in the discontinuous permafrost zone, pumping stations were required wherever there was even a small flow of water. Constant attention was required to prevent slush forming around the intake: pumps were not left unattended unless the pumping was continuous; if the pump shut down, or was shut down, because of lack of water, the pump lines were blown out immediately and all water removed so that the lines would not freeze solidly before the next pumping cycle began (Pike 1966).

The placing in stopes and raises of backfill from the surface was restricted to the summer season, i.e., when the material was thawed and could be handled easily. Once placed the permafrost was an advantage because the low temperature quickly froze the fill into a solid mass, permitting the removal of pillars with little danger soon after completion of the backfilling. The serpentinized peridotite, which tends to spall under normal conditions, could be held with a minimum of support. It was still necessary to backfill because of the flat dip of the ore, but this could be done after the ore was completely removed (Pike 1966; Weber and Teal 1959).

One of the largest mining ventures in the permafrost region of Canada is Pine Point Mines located at Pine Point, NWT, 50 miles (by railway) east of Hay River and eight miles south of Great Slave Lake. In late 1965 it began producing lead-zinc concentrates which are shipped south on the Great Slave Lake Railway. It is located in an area in the southern fringe of the discontinuous permafrost zone where permafrost occurs in scattered islands in peatlands but none has been found in the bedrock in which the ore is located. It has not been encountered at the townsite and thus is not a factor in the mining operation.

c / Silver

Echo Bay Mines began mining silver near Port Radium at the eastern end of Great Bear Lake in 1964. By 1965 it was mining ore with the highest silver content in Canada. No reports are available on the effects that permafrost has on the mining operation but permafrost conditions are presumably similar to those encountered at the Eldorado mine at Port Radium.

d / Tungsten

The first tungsten mine in Canada was opened by Canada Tungsten Mining Corporation Limited in 1964 in the southwest corner of Mackenzie District, near the Yukon border (White 1963), in the middle of the discontinuous permafrost zone. Permafrost was encountered in excavations to a depth of 35 feet. No reports have been received since 1964 but its location, about 200 miles north of the southern limit of permafrost and with a mean annual air temperature of 25° F, suggest that islands of permafrost 50 to 100 feet thick can be expected. Permafrost was encountered in excavations for mill foundations and in road construction, but there is no indication whether it is a factor in the mining operation.

e / Uranium

Mining of uranium began at Marian River, between Great Slave Lake and Great Bear Lake, in 1957 and closed down in the early 1960s because of ore depletion. It was located in the northern part of the discontinuous zone, where permafrost is widespread and generally about 200 feet thick. However, no reports of the effects of permafrost on the mining operation have been obtained. Conditions may have been similar to those encountered at Yellowknife although the average thickness of the permafrost is somewhat greater at Marian River.

3 *Provinces*

a / Base metals

The first base metals mine to operate in the permafrost region in the provinces was Sherritt-Gordon Mines Limited which began production in 1952 at Lynn Lake in northwestern Manitoba. The mean annual air temperature is about 25° F. It is located in the middle of the discontinuous zone and the permafrost is probably about sixty or seventy feet thick. Scattered islands of permafrost were encountered in the exploratory drilling and in the townsite. No difficulties have been caused by permafrost in the mining operations. In the late 1950s base metal mines were opened by Hudson Bay Mining and Smelting about fifty miles east of Flin Flon at Chisel Lake and Snow Lake. These mines are located near the southern limit of the permafrost region. Scattered islands of permafrost occur in peatlands in the vicinity but no permafrost has been found in the bedrock in which the mines are operating. At about the same time nickel mining began at Thompson, Manitoba. Permafrost has created problems in the townsite but it does not apparently affect the mining of ore.

b / Uranium

Mining began in 1948 on the north shore of Lake Athabaska and within a few years sixteen mines were operating in the Beaverlodge area which was served by the town of Uranium City; now only two mines, Gunnar and Ace, remain. This area is located in the Precambrian Shield in the northwest corner of Saskatchewan. Its position, in the middle of the discontinuous permafrost zone, is similar to Lynn Lake and Thompson. The mean annual air temperature is about 25° F and islands of permafrost up to sixty feet thick have been encountered (Pike 1966).

Permafrost has not caused extensive problems but it did affect one phase of construction. The main shaft of Eldorado Mining and Refining Co. Ltd. was constructed in 1951 near the centre of a basin-like depression in the glaciated rock hills. It was discovered that this basin was filled with thirty feet of frozen silt having a high ice content. Removal of the overlying peat caused the frozen silt to thaw into a slurry which created very unstable working conditions for installation of the shaft (see Figure 37).

The existence of permafrost was beneficial in the mining operation where frozen overburden formed a cap over the ore deposits since the ore was mined up to the frozen overburden from below. This allowed low cost recovery of ore under conditions that would otherwise have been very expensive. In some instances the permafrost retreated causing some minor slumping before the area was mined out and backfilled.

c / Asbestos

Mining began in northwestern British Columbia when the Cassiar Asbestos Corporation started production in 1955. An 86-mile road connects with the Alaska Highway to the north at Watson Lake, YT. Here the distribution of permafrost is greatly influenced by the mountainous relief – permafrost is not found in valley

FIGURE 37 Thawing of perennially frozen silt having high ice content at Eldorado Mine, Uranium City. Thawing was caused by removal of the natural vegetation followed by excavation.

bottoms but occurs at higher elevations and is more extensive on north-facing slopes than south-facing slopes. The orebody is located at an elevation of 6,000 to 6,500 feet above sea level, and a 3-mile-long tramline transports the ore to the mill located in the town at an elevation of 3,500 feet (see Figure 38).

No permafrost was encountered in the townsite during construction of the town or mine buildings. Up to 4,500 feet a few islands of permafrost were encountered in the excavations for the tramline footings. Above this, permafrost is widespread. The orebody lies entirely within the permafrost zone although it is on a south-facing slope. An exploratory adit was put in at the 5,700-foot level on this slope but no permafrost was encountered, but on the other side of the same mountain, on a north-facing slope, permafrost was located at an elevation of about 5,500 feet (R. J. E. Brown 1965c). Mining is open pit, the ore being mined in 30-foot benches. The buildings at the mine site are all constructed on permafrost, including one housing a furnace to thaw the frozen ore before it is transported on the tramline to the mill. In the mining operation permafrost has not

FIGURE 38 Tramline at Cassiar Asbestos Corporation mine, Cassiar, BC. There is no permafrost in foreground at bottom of tramline at an elevation of about 3,500 feet above sea level. Permafrost is widespread at top of tramline on mountain in background at the 6,000-foot elevation.

caused much difficulty, except that special measures are required to blast the frozen ore containing layers of ice.

d / Iron

Iron mining in the permafrost region of Canada is carried out entirely in Quebec and Labrador although large reserves of iron ore have been found in Baffin Island and Yukon Territory. The Labrador Trough is the largest single potential area of iron ore in Canada and the bulk of the country's present production is obtained from it. This area, 750 miles in length and approximately 50 miles in width, starts in the Ungava Bay area and extends southward past Schefferville and Wabush. There are indications that the Labrador Trough may continue northward into Baffin Island. The central portion in Quebec and Labrador, centred around Schefferville and developed by the Iron Ore Company of Canada in the early 1950s, is known to contain about 400 million tons of direct shipping ore grading about 52 per cent natural iron. Formations of concentrating-type ores around Labrador City and Wabush are known to contain several billion tons of reserves. In the southern part of the general area, at Gagnon, where Quebec Cartier Mining is operating, an aggregate of eleven billion tons of ore has been proven, with much additional tonnage inferred (Mining in Canada, 1967b).

Although exploitation of the vast iron ore reserves in the Labrador Trough did not begin until the mid-1950s the existence of this mineral wealth was known seventy-five years ago when A. P. Low, traversing this land in the 1890s for the Geological Survey of Canada, observed iron ore in the ridges and predicted that exploitation would begin within fifty years. These reports were disregarded until just prior to the Second World War when Hollinger Gold Mines Limited, which was searching for base metals in Ungava, discovered so much iron ore that an investigation was initiated using Low's accounts as the main source of information. In 1945 the first buildings were constructed on the shore of Knob Lake in the centre of the Labrador-Ungava Peninsula where the town of Schefferville (population approximately 5,000) now stands, and prospecting field parties operated from this base. By 1949 the enormous size of the project was seen to require large amounts of capital to carry it through to production. A new company, the Iron Ore Company of Canada, was formed and financed by Hollinger Gold Mines Limited, Labrador Mining and Exploration, M. A. Hanna Company of Cleveland, and several major American steel producers. Sufficient reserves of commercially profitable ore required to justify the monetary investment were proven in 1950, and plans for bringing it into immediate production went into operation (Humphreys 1958).

The town of Schefferville, the largest and oldest iron mining settlement in Labrador-Ungava, is 1,800 feet above sea level, and connected to Sept Iles on the north shore of the Gulf of St. Lawrence by a 360-mile-long railway line (the Quebec North Shore and Labrador Railway). Iron ore is shipped to tidewater on this railway. In the early 1960s the Iron Ore Company of Canada began producing iron concentrates and pellets at its Carol operation at Labrador City, 150 miles south of Schefferville. About the same time Wabush Mines began producing concentrates and pellets in the same area. Shortly thereafter Quebec Cartier Mining began production at Gagnon, 100 miles southwest of Labrador City and Wabush.

The latter two settlements, each having a population of several thousand, are connected by a thirty-mile branch line to the Quebec North Shore and Labrador Railway, by which the iron ore is shipped to Sept Iles. Gagnon is connected to the Gulf of St. Lawrence by a railway built by the mining company to Port Cartier, thirty miles west of Seven Islands. Ore is shipped by ocean freighter to the Eastern Seaboard, Great Lakes, and Europe. Shipments in 1967 were: 6.7 million tons of direct shipping ore from Schefferville; 6.5 million tons of concentrates and pellets from Labrador City; 6 million tons of concentrates and pellets from Wabush; and 8.3 million tons of concentrates and pellets from Gagnon (*Mining in Canada* November 1967).

The Labrador Trough is a belt of parallel ridges and valleys closely resembling the Appalachians. The folds were developed in a thick series of Proterozoic sediments and volcanics accumulated in a late Precambrian geosyncline that was later compressed from the northeast. The sediments, corrugated and in places overthrust by the compressive forces, are presently disposed in tight folds trending approximately north northwest – south southeast. Throughout the trough, the sediments are rich in iron formations (Hare 1952).

The climate in the Labrador Trough is subarctic with mean annual air temperatures ranging from about 29° F at Gagnon, to 24° F at Schefferville, to 20° F in the northern unexploited portion. It lies entirely within the discontinuous zone, the southern part in the southern fringe and the northern part in the subzone where permafrost is widespread. Permafrost has not been encountered at any of the above-mentioned townsites, but scattered islands occur in neighbouring peatlands. The hilly to mountainous relief of the Labrador Trough and other regions of Labrador-Ungava result in the occurrence of permafrost at higher elevations. This is the situation at Schefferville and Labrador City where extensive areas of the orebodies above the towns are perennially frozen to depths of 200 feet and more. No reports of permafrost occurrences have been recorded by Quebec Cartier Mining at Gagnon. Nor are reports available on the effects of permafrost at Wabush. Thus, the following description of the role of permafrost in iron mining in the Labrador Trough is restricted to the Iron Ore Company of Canada operation at Schefferville. Perennially frozen iron ore has been mined here for ten years and considerable experience has been gained by this company in coping with mining problems caused by permafrost.

Mining operations at Schefferville consist of four phases culminating in the shipment of ore by rail to Sept Iles. These operations – blasting, loading trucks, preliminary crushing, and loading ore trains – and the difficulties caused by permafrost are described in the paper by Ives (1962) from which the following account is taken. Two types of rigs are used for drilling blasting holes – churn drills were used initially but these have been superseded by rotary drills which bore a 9-inch diameter hole faster and more effectively than the former type. The blasted ore is loaded by power shovel into trucks each having a carrying capacity of 35 tons of ore (some larger trucks presently being tested have capacities up to 95 tons) which carry the ore to a screening and crushing plant located beside the railway. Regulations require that the ore be in chunks less than 4 inches in diameter. The ore is carried on a conveyor belt and dumped into hopper railway cars each capable of carrying 90 tons. Each ore train consists of three diesel locomotives and approximately 100 hopper cars.

Mining was initiated in 1954 and a token shipment of one and three-quarter million tons was taken from Ruth Lake mine. In 1955, however, full operations began when French and Gagnon mines were opened; production reached 10.5 million tons. During mining operations in this year, small bodies of perennially frozen ore were encountered in Ruth Lake mine and two of the Gagnon pits which caused some local handling difficulties. These occurrences of permafrost were a maximum of perhaps a few feet to several tens of feet thick, containing ice layers ¼ to ½ inch thick. These orebodies were close to the elevation of the town and the existence of permafrost was marginal.

Trenching operations in 1955 in the large Ferriman orebodies – north of Schefferville, at an elevation of 2,500 feet (900 feet above the town) – led to the discovery of permafrost a few feet beneath the ground surface. A vertical shaft driven into this orebody to a depth of 50 feet in 1948 encountered permafrost. Subsurface investigations in 1949 and 1950 at two orebodies 25 and 30 miles farther north also found permafrost.

The perennially frozen ore causes serious problems in the mining operation at several stages – drilling and blasting, removal of ore from pit face to railway car, and transportation by rail.

Blasting problems increase in proportion to the ice content of the ore. Where ice content exceeds 10 per cent by weight in the form of lenses and layers ¼ to 2 inches thick, a large proportion of the energy generated by each blast is absorbed. The result is incomplete and unsatisfactory rupture of the pit face. The frozen ore breaks into large, unmanageable blocks weighing up to 100 tons (see Figure 39), the success of each succeeding blast in the same area is reduced, and ore removal may be brought to a standstill. As the ideal angle of the ore face is reduced and the toe becomes more extended, so the possibility of a good blast becomes more remote. In practice, far greater amounts of explosives are required in permafrost than in unfrozen ground. If this larger quantity of explosion is used to blast unfrozen ore, which is thought mistakenly to be frozen, the resulting explosion causes large fragments of ore to be blown a considerable distance beyond the danger area, a great hazard to workers. As the toe becomes unmanageable secondary blasting becomes necessary, with accompanying stoppages in removal and loading of ore. Considerable research into types of explosives, method of installation, and spacing of charges has resulted in greater efficiency and reduction in the cost of mining.

The removal of ore from the pit face is closely associated with the success or failure of the blasting process. A typical blast in frozen ore results frequently in the production of large blocks, many of which cannot be handled by the electric shovels and trucks. In some instances the shovels are able to scrape off small fragments of frozen ore from the ore face but this is laborious and slow. Common practice has been to bulldoze large blocks to the centre of the pit and await natural thaw and disintegration, or to employ secondary blasting, or to attempt fragmentation of the blocks by mechanical percussion. This is accomplished by repeatedly dropping a pear-shaped steel ball (called a "frost ball") weighing several tons

FIGURE 39 Freshly blasted rocks of perennially frozen iron ore at Schefferville, PQ, in discontinuous permafrost zone. These blocks have to be broken into small fragments by mechanical means or left to thaw naturally before the ore can be trucked to the screening plant.

on the blocks from a crane. After three or four blows, the block is usually sufficiently fragmented to be loaded into the trucks. The disadvantage of the "frost
ball" is that only a small amount of frozen ore can be handled at a time.

If the supply of large blocks is heavy, the pit floor becomes congested – increasing the difficulty of fully effective deployment of the machinery and, accordingly,
reducing production. Another problem is that secondary blasting of the toe of the
ore face, or only partially successful primary blasting, tends to result in an uneven
pit floor. An extreme case is when the shovel cannot approach sufficiently close to
reach ore from the upper part of the pit face.

Successful breakdown following primary blasting, but still leaving frozen
fragments exceeding one to two feet in diameter permits transfer of the frozen ore
from the pit to the screening plant. Sometimes the fragments re-freeze into a mass
in the dumper of the truck by the time it has arrived at the screening and crushing
plant. Also, the crushing of frozen fragments larger than the maximum four-inch
screening size, is more difficult than crushing unfrozen pieces. Increased wear and
tear on the plant, jams and further reduction of production, and increase in cost
result from the difficulties described above.

Transportation to Sept Iles on warm summer days leads to the gradual thawing
of at least part of the frozen ore. This can accentuate the problem of so-called
"sticky ore," i.e., when fine-grained ores with a critical moisture content partially
stick to the sides and bottom of the railway cars during dumping at the Sept Iles
terminal. Continued thawing presumably extends this problem to all transport
operations until the blast furnace is reached.

Because of the importance of knowing the distribution of permafrost in the
orebodies and the opportunity of improving knowledge of the occurrence of
permafrost in Labrador-Ungava, a co-ordinated research programme was initiated
in 1959 by the Iron Ore Company of Canada, the McGill Subarctic Research
Laboratory, and the Division of Building Research of the National Research
Council. Detailed ground temperature measurements have been taken monthly
since 1959 on thermocouple cables installed in a test area adjacent to the Ferriman mine. Investigations of climate and terrain, involving vegetation, snow cover,
and ground surface characteristics have been made during this period. More work
remains to be done on the data before final conclusions can be drawn, but some
broad correlations relating permafrost and ore distribution appear to exist.

Over wide areas the permafrost appears to be a product of the present and
recent climate, exceeding 150 feet in thickness in many places and probably
occurring in thicknesses exceeding 300 feet. There is a strong correlation between
vegetation type, depth of snow cover, and development of permafrost (Annersten 1964). It is suggested that all areas extending above the treeline and south
to the Laurentian Scarp, which is just north of the Gulf of St. Lawrence, and

especially those where the wind prohibits the accumulation of deep snow, are areas of potential permafrost. Below the treeline it is possible that patches of permafrost formed in past cooler climates may occur across wide areas of central and south-central Labrador-Ungava.

Virtually all of the untapped orebodies of the Iron Ore Company of Canada are located north of Schefferville at higher elevations than the town of Schefferville and the orebodies being mined at present. Permafrost becomes increasingly widespread and thicker toward the north and will be an increasingly important factor in the cost of mining. The railway has already been extended several miles north of Schefferville to reach new orebodies that will soon be coming into production and further extensions will take place in the future.

Research on improving the methods of predicting the distribution and temperature regime of permafrost in the orebodies and on improving methods of mining the frozen ore continues (Thom 1969). Attempts are being made to delineate bodies of permafrost prior to mining by geophysical (seismic and resistivity) methods, shallow trenching, drilling, installation of thermocouple cables for direct temperature measurements, surveys of snow conditions, vegetation and drainage, and study of microrelief including patterned ground. Long-range investigations on conditions of ice within various rock types, and the possibility of modifying ground temperatures by changes to surface conditions (e.g., increase snow cover by snow fences), are being considered in future research programmes. The three main problems in mining frozen ore, those of blasting, removal of ore from pit face, and transportation to the south, are continually under consideration.

C POTENTIAL DEVELOPMENTS

1 *Yukon Territory*

a / Base metals

Anvil Mining Corporation is developing a large lead-zinc mine in the central Yukon Territory about 120 miles north of Whitehorse and the same distance east of Carmacks. The company holds 130,000 acres in the area on which some 50 million tons of lead-zinc ore have been proven with some silver. An additional 25 million tons have been proven in the area by other companies and it is expected that the total ore in the region may amount to some 100 million tons, making it a major lead-zinc area in world terms (*Mining in Canada* 1967c).

The mine is located in the southern fringe of the discontinuous permafrost zone where the mean annual air temperature is about 26° F. Permafrost varying in thickness from 15 to 60 feet has been encountered in the proposed townsite. Soils vary from granular coarse-grained sands and gravels containing little or no ice to fine-grained silts with ice lenses up to ¾ inch thick. The hilly to mountainous relief will probably mean that greater thicknesses of permafrost will be encountered in mining at elevations above the townsite and on north-facing slopes. Permafrost has been encountered in exploratory trenching and drilling and may be an important factor in future mining operations (Aho 1966).

b / Iron

A large hematite deposit extending over several hundred square miles was discovered in 1961 near the Yukon-Northwest Territories border a short distance south of the Arctic Circle, approximately 200 miles east of Dawson. Crest Exploration Limited made the discovery and has staked claims whose reserves are estimated in excess of 11 billion tons of hematite grading from 45 to 48 per cent iron. The reserves include 2 billion tons of open pit material requiring concentration to over 60 per cent iron before it can be marketed. Long-term markets for 5 to 10 million tons per year would be necessary to justify the large capital investment required for the plants and transportation facilities (Dubnie and Buck 1965, *Mining in Canada* 1967*b*, *Northern Miner* 1962).

This area is located in the middle of the discontinuous zone; permafrost is 200 feet thick at Dawson and 450 feet thick in the mines at United Keno. Thus, it appears that permafrost of this order of thickness exists in the Crest area. No report is available of permafrost conditions in the area or the effects that it would have on mining; however, it seems probable that it would present problems similar to those encountered at Schefferville. It could also be a factor in the construction of buildings and roads, and of the railway proposed for transporting the concentrated ore from the mine. The development of this resource will depend on availability of markets.

c / Oil

Little information is available on oil potential in Yukon Territory but exploratory drilling has been carried out at two widely separated locations. The first is in the Hyland Plateau in the extreme southeast corner of the Territory. No information has been recorded on permafrost conditions in this area but its location in the southern fringe of the discontinuous zone, where permafrost exists mostly as scattered islands in peatlands suggests that permafrost may not be a major factor in the development of oil reserves in this region.

The second area of oil exploration is in the Peel Plateau where conditions are much more severe than in the Hyland Plateau and permafrost exceeding 1,000 feet in thickness has been encountered. Several holes were drilled by Western Minerals Limited in 1959 and 1960. During the years 1963 to 1965 Mobil Oil Canada Limited drilled ten wells in this area (Dier 1969). Two problems associated with the drilling operation were obtaining a stable base for the drill rig and associated equipment, and installing casing in the permafrost layer. The first problem was solved by the use of wood piles frozen into drilled holes or where possible the use of gravel or fragmented shale pads placed on the moss-covered ground surface. The problem of drilling and casing in the permafrost was solved by the use of air as a circulating medium while drilling and the use of special techniques in the formulation, handling, and placing of the cement.

2 *Northwest Territories*

a / Copper
Active staking and exploration work is currently underway in the Coppermine, NWT, area by Coppermine River Limited, a company newly formed by five established mining companies. High-grade copper minerals (bornite – about 40 per cent copper, and chalcocite – about 79 per cent copper) have been found in the area where Eskimos used native copper 200 years ago. There is no indication whether this area will be developed but a firm price has been obtained from a Danish shipping firm for the transport of copper concentrate to Japan, via Bering Strait, should production emerge from the project (*Northern Miner* 1967).

Coppermine is located in the southern part of the continuous permafrost zone making it one of the few active mining exploration projects in this zone. The thickness of permafrost is not known but it probably exceeds 500 feet and could be an important factor in mining and in the construction of buildings and roads.

b / Iron
The discovery of a huge and rich deposit of iron ore in northern Baffin Island in 1962 prompted several mining companies to form Baffinland Iron Mines Limited which has been carrying out an active exploration programme and feasibility studies since that time. Potential high-grade reserves are reported to be in excess of 100 million tons of open pit material grading 68–9 per cent iron in four separate zones. This average ore grade is the highest recorded in Canada on any large deposit. It appears also that the ore reserves greatly exceed the amount presently proven by drilling.

The deposits occur in sedimentary iron formations in a complex of Precambrian metamorphic and igneous rocks. The iron minerals occur as fine-grained blue hematite, magnetite, and specularite, all high grade. Low-grade ores also occur (Dubnie and Buck 1965). These high-grade orebodies are located in the northern part of the continuous zone where climatic and permafrost conditions are very severe. The mean annual air temperature is about 7° F. Permafrost occurs everywhere beneath the ground surface and probably exceeds 1,000 feet in thickness.

Present development plans call for the construction of a town site and mining facilities at the mine site, a sixty-five-mile railway to tidewater at Milne Inlet, and another town site and loading facilities at sea level. Permafrost problems similar to those at Schefferville would probably be encountered. One difference is that the permafrost temperatures are much lower in the Baffinland development – 10° F at the 50-foot depth, in contrast to about 29° F at Schefferville. Thus, frozen ore containing ice may be more difficult to blast and mine. Preliminary drilling in the town site area at the mine site has shown up the presence of large masses of ground ice. Extensive thawing of the permafrost due to construction and operation of buildings could lead to serious problems. Fortunately, unlimited amounts of gravel are available in the area (Watts and Megill 1966, Samson 1969).

It is not known if and when production will begin in this area. The ore reserves rank among the richest and largest in the world but the factors at present critical in determining future moves are the shipping season of only six weeks and long distance to potential markets.

c / Oil

The oil field at Norman Wells, NWT, on the Mackenzie River is the only producer in the permafrost region of Canada. In the past ten to fifteen years, it became evident through geological exploration that potential petroleum reserves do occur in northern Canada. In the Northwest Territories a strip 200 to 400 miles wide down the Mackenzie Valley – reaching from the sixtieth parallel to the Arctic Ocean – contains about 200,000 square miles of sedimentary area, with an estimated average thickness exceeding 7,000 feet. It appears that potential petroleum reserves could be 8 to 13 billion barrels and potential natural gas reserves 50 to 80 trillion cubic feet (Nickle 1961).

The oil potential in the Arctic Archipelago was publicized mainly by operation "Franklin," a major survey programme carried out in 1955 by the Geological Survey of Canada, which produced information on the geology of the great sedimentary formations existing there (Dubnie and Buck 1965). The Arctic Islands are estimated to contain about 78,000 square miles of sedimentary area, with an average depth of as much as 15,000 feet. It appears that potential petroleum reserves could be 5 to 7 billion barrels and potential natural gas reserves 30 to 45 trillion cubic feet (Nickle 1961).

In the past decade considerable exploratory work has been carried out in the Mackenzie River valley and Arctic Archipelago. Initial impetus to private activity was provided in 1959 when large tracts of land were opened for exploration and nearly 50 million acres were taken up by private companies. At the end of 1964 approximately 60 million acres were held under permit on the territorial mainland (Yukon Territory and Northwest Territories, but mostly the latter) and approximately 50 million acres were held in the Arctic Islands. By 1965 over 50 million acres were taken up in offshore rights in Hudson Bay and on the arctic coast of the mainland.

More than one hundred exploratory wells have been drilled, together with extensive airborne geological and geophysical reconnaissance, surface geological work, and ground seismological work. Exploration in the Arctic Islands is at an earlier stage. Considerable airborne and ground geological work has been done, initially by the Geological Survey of Canada and later by a number of oil companies. A number of very large structural features – the sort that could yield major oil or gas fields – have been surface mapped.

The three wells which have been drilled in the Arctic Islands in the 1960s attracted considerable publicity because they were the first to establish the feasibility of drilling for petroleum in high arctic conditions (Personal communication, Department of Indian Affairs and Northern Development). The first was drilled in 1961–2 by Dome Petroleum Company and several other small oil companies at Winter Harbour on Melville Island to a depth of 12,500 feet at a cost of about two million dollars (Armstrong 1961); a second was drilled by Lobitos Company (now part of Great Plains Company) in 1963 on Cornwallis Island to a depth of about 4,800 feet at a cost of 1.5 million dollars; a third was drilled by Dominion Exploration Company in 1964 on Bathurst Island to about 10,000 feet at a cost of two million dollars. The last which is located about 950 miles south of the North Pole is the most northerly well ever drilled. At none of the three wells was petroleum encountered. Private companies are continuing, however, to take out acreage and to explore with a view to drilling in the future (Dubnie and Buck 1965). Late in 1967 the Federal Government announced the establishment of Panarctic Oils Limited, composed of a consortium of several Canadian oil and mining companies with participation by the government as the biggest single shareholder, to carry out an oil exploration programme in the Arctic Archipelago. Two holes were to be drilled in the spring of 1969 on Melville Island.

The distribution of permafrost in the potential petroleum areas described above varies from its appearance in scattered islands in the southern fringe of the discontinuous zone at the sixtieth parallel to its continuous presence more than 1,000 feet thick in the Arctic Islands. A 2,000-foot thermistor cable to measure ground temperatures was placed in the hole on Melville Island. There temperatures below 32° F were observed to a depth of 1,500 feet, indicating that permafrost extends at least to that thickness.

Permafrost may cause some special problems in both the exploration and production phases of oil exploitation. Some, particularly in the exploration phase, are similar to those problems encountered in the exploration of mineral deposits.

Others, particularly in the production phase, are peculiar to oil deposits for two reasons: because oil is a liquid, and because the oil is deep-seated. Usually the petroliferous deposits are situated below the permafrost layer, in contrast to the solid-state minerals which are perennially frozen. In the case of oil the problems consist of penetrating the frozen zone of the earth's crust with drilling equipment and maintaining a flow of oil upwards through this zone of below freezing temperatures to the surface.

Geophysical prospecting of petroliferous deposits is affected by permafrost, particularly where it extends downward for hundreds of feet. Prospecting by seismic or electrical resistivity methods requires a knowledge of permafrost distribution because seismic wave velocities are nearly doubled in frozen ground and the electrical resistivity of these rocks is much higher than of similar unfrozen strata. In some instances this has resulted in data that were interpreted as indicating bedrock structures where none exist. This occurred, for example, in the major oil exploration programme in Northern Alaska from 1944 to 1953 largely within Naval Petroleum Reserve No. 4, abbreviated to Pet 4 (Reed 1969). In the seismic investigations, all horizons showed an elongate syncline under a lake on the coastal plain sufficiently large and deep to prevent its freezing to the bottom in winter. After considerable study it was concluded that the apparent syncline was the result of decreased velocities in unfrozen material beneath the lake and not a structural feature.

Another critical problem in oil exploration, which will require careful consideration as it increases in the Arctic, is the thermal erosion of seismograph lines bull-dozed across the tundra through areas of high ground ice content. Such difficulties have arisen in the Tuktoyaktuk and Fort McPherson areas (Watmore 1969). In the former area surface vegetation and soils were peeled back by bulldozer in October, 1965 to a width of more than 14 feet and a depth of approximately 10 inches leaving a relatively smooth firm surface near the permafrost table on which geophones could be placed. Three years later, thawing and settlement of the exposed permafrost produced a lengthy ditch-like depression exceeding a depth of 6 feet in places where ground ice content was particularly high. On the Peel Plateau near Fort McPherson a gulley developed along a cutline which reached a width of 23 feet and a depth of 8 feet at one location in 4 years. This gulley extended up a 3 per cent slope. Snow accumulation in the gulley accentuated spring runoff, which hastened erosion, and a steady flow of water from melting ground ice occurred even during dry periods in summer.

After geophysical prospecting has outlined oil deposits of sufficient magnitude to warrant exploratory drilling, other problems appear. Well-drilling requires modifications of standard techniques used in temperate climates. Difficulties are encountered often in obtaining proper foundations for the drill rigs, particularly if the unconsolidated overburden has a high ice content. The main precaution is to prevent the transmission of heat from the engines of the drill rig and the drilling mud pits to the ground especially in summer since this will thaw the ice and cause excessive settlement of the soils. In winter a timber mat lying on the undisturbed ground vegetative cover provides an adequate foundation, but for summer drilling the rig and mud pits may have to be placed on piles driven to varying depths depending on the load each one must sustain. To protect the surface soil from thawing and severe settlement, refrigerating liquid can be circulated through a piping system under the timber mat, but this adds to the drilling costs. When the rig is to be moved, steam instead of refrigerant can be circulated through the pipes in order to thaw the frozen soil adhering to the timbers and free them for easy removal (Vonder 1953). Various foundation designs, depending on the estimated load, the length of time the hole was expected to be drilling and the season, were used for drill rigs in the Pet 4 project (Reed 1969). Rigs on frozen silt with high ice content were placed on piles frozen into the silt. Refrigerated foundations were used in several cases with satisfactory results, for example, at one site where drilling was expected to continue for two years.

After the rig and drilling mud pits have been anchored firmly, difficulties may arise in the drilling of the hole. The perennially frozen state of the unconsolidated overburden and underlying rock does not affect the drilling time appreciably, but circulation of drilling mud must be maintained at all times to prevent the drill rod freezing in the hole and the possible loss of equipment. Difficulty may be experienced in keeping the drilling mud at the proper temperature and finding an adequate water supply or proper local material for the mud. In shallow holes particularly, the tools will freeze in after a few hours of idleness. Steam may be used to maintain the mud temperature at about 60° F (Fagin 1947). If the mud is much warmer, when it is pumped down through the drill rod it may thaw the wall of the hole and cause excessive caving. On the other hand, deep wells extending far below the permafrost may encounter temperatures as high as 100° F to 150° F. Drilling mud is warmed sufficiently to thaw the permafrost in the vicinity of the hole on its return to the surface, creating the possibilities of caving of the hole and severe settlement of the rig foundation (Black 1956).

After the hole is drilled, several difficulties are encountered in obtaining an effective cement job around the casing that is set in permafrost. The cement sometimes freezes before it is properly placed or it may emit so much heat while it is setting that it thaws the wall of the hole and fails to provide a good bond between the rock and the pipe (Vonder 1953). Cement has often failed to set in perennially frozen strata, and even below the frozen zone where the temperature is below 40° F the setting is retarded. It has been found that these difficulties can be overcome partly by quick-setting cements, heating the mixing water, and circulating warm mud while the cement is setting (Brown and Wopnford 1955). Recent experiments indicate that Portland cements (especially high early strength cement) can be used successfully if precautions are observed and proper techniques are used. The slurry temperature should be maintained at about 60° F and the cement heated by some external source, such as warm water following the plug of warm fluid circulated in the casing after the placing of the cement in the annulus (i.e., the circular space in the hole between the casing and the rock). High alumina cement appears to be most suitable. It will set and gain adequate strength even without any external heat (Cameron 1969).

Even after the cement has set, the permafrost may cause the casing to collapse. This is usually attributed to the freezing of the drilling mud in the annular space or annulus, resulting in tremendous pressure, up to 10,000 pounds per square inch, on the casing. Even if the casing has collapsed there is little danger as long as the ice remains frozen, but if production is obtained at depth, the warm petroleum passing upwards will melt the ice and cause pressure on the wall of the hole. In this way, the oil field as well as the well itself could be easily damaged (Vonder 1953).

To overcome such a possibility, it is necessary to prevent the formation of ice by one of two methods: the first is to lower the freezing point of the mud in the annulus with brine so that ice will not form, but unfortunately it is sometimes difficult to hold the required concentration of brine for a sufficient period; the second and more usual method is to replace the mud in the annulus with some fluid such as oil which will not freeze (Vonder 1953).

In Alaska, oil-bearing sands were discovered at depths in the perennially frozen zone. Besides the problem of obtaining a good set for cement, there is the added one of possible blocking of the sand pores by ice formation. If the connate water is highly saline it probably will not be frozen, but any infiltration of fresh water from the drilling fluid will reduce the salinity and permit ice formation at higher temperatures (Vonder 1953).

In addition, if the permafrost is acting as a trap for the oil or even contains oil reserves, the low temperatures adversely affect the asphalt base types particularly. This increases production difficulties, decreases yield, and increases costs (Black 1956).

Although hydrate formation in producing wells is common in temperate zones, the low temperatures in permafrost make their formation inevitable. This condition can be overcome by installing the same type of resistance heaters used to prevent ice formation in the tubing of shallow oil wells although this results in loss of time. It is sometimes necessary to circulate a hot brine solution or heated oil through the drill rod in the tubing (Vonder 1953).

Experiences in the USSR are also relevant to the exploitation of potential petroleum reserves in the Canadian Arctic. Oil was discovered at Nordvik in

northwestern Siberia where the permafrost is 2,000 feet thick (*Atomes*, 1954).
The oil was formed long before the formation of the permafrost, the latter forming
an impervious capping above the oil and so preventing it from rising.

Because of the close proximity of the oil to the permafrost, the temperature of
the oil deposits was near 32° F. When the caprock became perennially frozen, the
amount of free gas in the oil decreased because of the increased solubility of the
gas in the oil as the temperature of the latter decreased. This is a handicap to
oil exploration because free gas is a prime force in moving the oil to the wells. The
oil is also more viscous at lower temperatures, and paraffin crystallizes in the pores
of the oil-bearing strata, greatly reducing the porosity of the rocks and the ease
of extracting the oil. If the temperature of an oil deposit should decrease to 32° F,
water under hydrostatic pressure, which is ordinarily another factor contributing
to the movement of the oil to the wells will freeze and the extraction of the oil is
handicapped. On the other hand, the imperviousness of the perennially frozen
caprock serves as a protection against de-gasification and prevents the possible
migration of oil away from the field. Difficulties have been reported in obtaining
good cementing jobs in the wells because of low temperatures. Thermal treatment
of holes by injecting gases to increase viscosity and decrease clogging have been
used with sufficient success to maintain production (Petrov and Rakitov 1940).

3 *Provinces*

a / Asbestos

In 1957 a large deposit of asbestos was discovered by Murray Watts in the northwestern extremity of Quebec. During the following four years diamond drilling
in the area, named Asbestos Hill, outlined 20 million tons. During the same period
a feasibility study was carried out for the construction of a town site and mill at
the mine 40 miles southeast of Deception Bay on Hudson Strait. A road is planned
from the mine to tidewater and port facilities will be developed on Deception Bay.
The Asbestos Corporation took over the option from Murray Mining Corporation
in 1964. Delays in development have occurred because of the difficulties in obtaining reliable markets but it is hoped that production will begin in the early 1970s
(*Northern Miner* 1966).

The mine is located in the southern part of the continuous permafrost zone.
The mean annual air temperature is about 17° F and it is estimated from drilling
that permafrost exceeds 900 feet in thickness. It is located about 300 miles north
of the treeline. The terrain is typical of the Precambrian Shield, a windswept rocky
upland with shallow cover of glacial drift and rock detritus. It is cut by winding
canyons so that in places local relief exceeds 1,000 feet (Lawrence and Pihlainen
1963).

It is not certain how extensive the effects of permafrost will be on mining the asbestos. The serpentine and chrysotile rock contains considerable quantities of ice particularly in the top layer which is severely weathered. Problems similar to those encountered in the iron mines at Schefferville could be experienced here. The nature of the top layer of rock has produced some difficulty in the foundation design for the mill and associated buildings which impose heavy loads on the underlying ground. The top ten or more feet of bedrock do not present firm bearing because this layer is badly fractured and contains large quantities of ice (Samson 1969). One proposed design is to excavate the overburden and place the plant on 39-inch diameter caissons drilled fifteen feet into the bedrock. Smaller and lighter one- and two-storey buildings will be placed on gravel pads with air spaces beneath the buildings to reduce heat loss to the perennially frozen ground (*Heavy Construction News* 1964).

An unusual feature of the development will be the water supply which will come from a reservoir created by a 200-foot long by 50-foot high dam built across a small stream. The dam will be constructed during the winter in layers; each layer will be frozen before the next layer is applied. Ground temperatures will be taken in the core and it is estimated that it will require eight years for the thawing effect of the water on the upstream side to reach the frozen core. As thawing occurs, the dam can be extended downstream by adding mine waste which will be available in large quantities. This design is considerably less expensive than a conventional earth fill concrete centre core structure of the same size (*Heavy Construction News* 1964).

b / Other minerals

Active exploration for minerals and oil is underway in the northern areas of all the provinces. Many mineralized zones are known in northern British Columbia, and exploitation of them may encounter permafrost problems similar to those at Cassiar. Mineral and oil exploitation in northern Alberta and Saskatchewan may not be affected as much by permafrost because conditions are not as severe. The permafrost region extends farther south in Manitoba and Ontario and permafrost could be more of a factor. Possible oil exploration in northern Ontario, for example, may encounter some permafrost problems.

In the provinces, permafrost probably exerts its greatest influence in Quebec and Newfoundland (Labrador). Iron mining at Schefferville has already experienced considerable difficulties and they will increase in magnitude as new mines are developed farther north in the Labrador Trough. In 1963 plans were underway for iron mining to begin at Hopes Advance Bay situated on Ungava Bay, at the north end of the Labrador Trough. By 1968 production was going to reach five million tons of pelletized iron ore (Guimond 1958a, b). This area is located in the northern part of the discontinuous zone and permafrost apparently is 500 to 1,000 feet thick. The mine did not go into production, not because of the permafrost but because markets could not be obtained for the ore. Nevertheless, permafrost no doubt will be a formidable problem in extraction of the ore.

Many mineralized zones occur in Labrador-Ungava outside of the Labrador Trough. Although none has been brought into production, many companies are carrying out exploration. It is probable that new mines will develop in the future and permafrost will present problems.

D   CONCLUSION

In reviewing the effects that permafrost has on mining activities in northern Canada, it is evident that it imposes several difficulties and few benefits. Problems vary depending on the type of mineral being exploited and the nature of the perennially frozen ground. With any given mineral, the magnitude of the problem generally increases toward the north as the permafrost changes from discontinuous to continuous. In the discontinuous zone permafrost does not occur throughout a mineral deposit and is not a problem throughout the mining operation. Therefor, standard techniques used in non-permafrost areas may be employed in mining areas not affected. If the distribution of permafrost is not definitely known, its unexpected presence may slow down the mining operation and upset production schedules. Where permafrost is widespread or continuous the difficulties it imposes are taken into account from the beginning of the operation and radical changes in production planning do not occur. On the other hand, however, perennially frozen mineral deposits require expensive and time-consuming modifications of standard procedures which may result in higher production costs and lower output.

Mining by open pit methods imposes somewhat different problems from underground workings. The frozen gold-bearing placers in the Klondike and the frozen iron ore deposits in Labrador-Ungava require thawing to facilitate extraction. On the other hand, subsurface workings often require that the underground temperatures be kept below freezing to prevent thawing and subsequent caving and flooding of mine drifts and galleries. Oil deposits are in a separate category because of their liquid character and their location, generally below the perennially frozen zone rather than in it.

These problems encountered in Canada's permafrost region can be surmounted by the various techniques devised to overcome them as is shown by the fact that the mineral resources are being tapped and marketed. Technically the problems are solvable, but economically it is not always feasible when these deposits have to compete with similar deposits in more southerly areas. In Canada most of the development to date has been in the southern reaches of the permafrost region where access is not as difficult as further north. Increasing attention is being paid to potential mineral resources in the high Arctic, where permafrost is continuous and conditions severe. Here there is more of a problem than in the discontinuous zone but permafrost will hamper mining activities rather than prevent them. The economic limitations on this activity in northern Canada will continue to be due to severe climate, difficulty of access, and great distance to markets.

8 Agriculture

Agriculture plays a minor role in Canada's permafrost region compared to other human activities. Through the years of development of this region, and particularly since the Second World War, the amount of money and effort expended in construction and mining has amounted to many millions of dollars and man-hours. New towns have been built, transportation routes have been established, and new mines have been brought into production. These various endeavours add to a sum total of activity that is growing at an accelerating rate. On the other hand, agriculture, which has been marginal from the earliest years of development of the Canadian North, has remained thus in the face of the continued boom of other activities. There are several aspects of this situation which illustrate the difference between agriculture and the other pursuits mentioned.

Agricultural products produced in this marginal region are consumed exclusively by the local population. Self-sufficiency is far from being a fact, however, because agriculture is almost exclusively of the garden plot variety – to supplement the large quantities of agricultural products that must be imported from the south to support the meagre and scattered population.

The number of people supporting themselves in agricultural pursuits even in the northern parts of the provinces within the permafrost region is very small – much less than the proportion dependent on agriculture in southern Canada. Nevertheless, there are localities which are suitable for agriculture in northern Canada. The deterrents at the present time are insufficient population to farm these lands and lack of markets for the produce. Once the effort is made to put such land into cultivation, it is possible that the present population could be supported almost exclusively by these crops.

Although agriculture is very limited in Canada's permafrost region, it is fairly extensive in Alaska and widespread in Siberia. In Canada, less than one per cent of the population lives in the permafrost region. Alaska, which also has permafrost, has less than one per cent of the total population of the United States. There are, however, about 225,000 people living in Alaska in an area of 571,065 square miles or 0.4 persons per square mile, which is more than ten times the density of the Yukon and Northwest Territories. For the USSR, exact population figures are not available, but there are probably 4,000,000 or more people occupying the 4.5 million square miles of permafrost area giving a population density of about one person per square mile, which is two and a half times the density of Alaska.

Although agriculture in Canada's permafrost region is presently of minor importance relative to other activities, it is possible that it may expand, particularly in Mackenzie District and Yukon Territory where physical conditions are most favourable. Agriculture is more developed in Alaska and Siberia because of higher population densities. In Siberia Russian policy has been aimed at developing local agricultural resources with considerable success. Physical conditions are an important factor in this development – summers are slightly warmer and growing seasons longer in Siberia than in northern Canada; soils are more suitable for cultivation and more widespread in Siberia because most of the region was not glaciated, in contrast to northern Canada where much of the soil, particularly in the Precambrian Shield, was removed by Pleistocene glaciers and ice caps.

Because of the relevance of long-standing Russian experience and, to a lesser degree, American experience to possible future agricultural developments in northern Canada, a brief description of the situation in these other two countries is included where pertinent in this chapter.

Many factors influence crop production in the permafrost region. Probably the most important is climate, of which the most significant aspects are the number of frost free days and the number of days above 42° F – the minimum limit for vegetative growth. In Canada and Alaska, and particularly in the USSR, these aspects have been the limiting ones in the attempts to extend agriculture northward. The length of daylight in summer partially offsets these growth limitations, promoting relatively fast maturation of plants. Precipitation, which is generally quite low in the permafrost region, is offset by the usual abundance of water in the seasonally thawed layer above the permafrost. Other factors such as evaporation have some effect.

With regard to the soil factor, it is important to note that a major difference between agriculture and other activities described in this book is the depth of ground affected and the definition of the term "soil." The various structures involved in the development of settlements, transportation, and mining – buildings, roads, runways, mine shafts, etc. – are all engineering structures affecting the thermal and moisture regime of the unconsolidated overburden and bedrock. The term "soil" includes the unconsolidated portion of the earth's crust lying on the bedrock. On the other hand, the agriculturalist in permafrost regions as elsewhere is primarily concerned with the top weathered portion of the unconsolidated overburden. The biological aspect is a problem of great importance – both the regime of the living organisms connected with crop growth and the nutritive regime.

The severe climate in the permafrost region inhibits soil profile development.

Profiles are best developed in river valleys in the southern fringe of the permafrost region where the climate is relatively more favourable and where deep deposits of fine-grained sediment have accumulated. Generally speaking, however, profile development is inhibited by the short frost free period and slow breakdown of organic material resulting in generally high acidity. The moisture and nutritive regimes of the soil and bacterial action are generally retarded.

Basically, the elements of the physical environment are mostly detrimental – permafrost, the result of a severe climate, is a handicap – but there are some beneficial and therefore compensating factors. Agriculture is possible, however, in more of Canada's permafrost zone than is now cultivated. Where there is arable land south of the northern limit of crop growth, advances in the development of faster maturing species are being made. Greenhouse cultivation is being conducted on a limited basis and could be enlarged. But economic factors enter the picture. Population is very small and demand is not high. The cost and effort of raising agriculture from the present garden plot type to a cash basis exceeds the cost of bringing in produce from the south. Nevertheless, the trend may develop for agriculture to expand in Canada's permafrost region as the population increases.

A NATURE OF SOILS

Less is known about the polar soils than of any other natural region in the world. Despite their low agricultural potential, a few soil scientists have worked in northern Canada and Alaska to classify them. In Canada the Department of Agriculture has conducted soil surveys in the Mackenzie District and Yukon Territory (Day 1963, Day and Rice 1964). Considerable pedological work has been carried out in northern Alaska (Tedrow and Brown 1967).

1 *Distinguishing Characteristics*

Distinguishing characteristics of soils in the permafrost region are their poor profile development and low temperatures, and the large accumulation of organic material particularly in the top horizons. The vegetation of the Subarctic and even the ground vegetation of the tundra is quite luxuriant. Because of the short growing season and low temperatures even at this time, however, bacterial activity is low and decomposition of organic material is slow. Even the mineral soils, therefore, have organic surface layers of varying thickness and the division of soils into the two categories of "mineral" and "organic" must be somewhat arbitrary. Usually, Canadian pedologists group soils having less than one foot of organic material above the underlying mineral soil in the "mineral" group, and those with one foot or more of surface organic material in the "organic" group.

2 *Zonal Soils*

Mineral zonal soils that have formed under relatively favourable climatic and vegetative conditions, such as those in the Mackenzie and Yukon River valleys, have little profile development, frequently less than one foot. Generally, there is a fairly abrupt division between the top organic layer and the mineral soil, and the weathered mineral horizon and mineral parent material. Leaching of calcium carbonate is slight. The differences among soils are mainly in the parent material, but topographic position also plays a role in that it affects the thickness of the surface organic layer. Soils on the level and in depressions usually have thicker organic layers than soils on slopes because of poorer drainage and aeration, and the slower rate or lack of decomposition (Leahey 1954, Day 1963, Day and Rice 1964).

Tundra soils are similar to the subarctic mineral and organic soils except that their profile is less well developed. Podzolization is relatively inactive in soils where permafrost is near the surface. There is generally a tough, fibrous, brown mat on the surface underlain by a few inches of dark-coloured, humus-rich soil. This fades to lighter coloured grey or mottled soil beneath, down to the permafrost table or to the unaltered parent material. In summer the soils are almost always either wet or moist beneath the surface and often are wet at the surface. Because of the low temperatures, physical weathering processes predominate, and chemical and biological processes are retarded. Growth is slow, but decomposition of plant remains by micro-organisms is slower, resulting in an accumulation of organic material at the surface. Some of this becomes mixed into the lower soil because of overturning by freezing and thawing. The clay content is relatively low. Partly because of the long periods of freezing and waterlogging, little leaching or eluviation takes place, although materials become mixed because of pressures on viscous soil during freezing (Kellogg and Nygard 1951, Day 1963).

3 *Azonal Soils*

Mineral azonal soils include those recent alluvial soils found along rivers and streams in northern Canada. These soils show no development in profile characteristics except for an accumulation of organic material in the upper part of the section (Leahey 1954).

In the case of organic azonal soils, no striking difference has been noted between arctic samples and those from temperate regions. Those in forested parts of the permafrost region do not differ much from others south of the permafrost region. Most of the organic deposits appear to be derived from mosses, often with an admixture of woody peat and sometimes sedge peat. The permafrost table lies usually 1½ to 2 feet below the ground surface.

4 *Factors Affecting Soil Development*

Much of the pioneer work on soils in the Canadian Arctic and Subarctic was carried out by A. Leahey of the Canadian Department of Agriculture. The following accounts of factors affecting soil development and the distribution of soils in these northern regions were derived mainly from his reports (Leahey 1943, 1944, 1953, 1954).

Wide variations in climate exist in the permafrost region of Canada. The forested Mackenzie and Yukon River valleys in the northwest experience fairly warm summers and very cold winters, but the tundra in the central part of northern Canada and in the northeast has cool summers and very cold winters. Precipitation varies, being generally greater in the Subarctic than in the Arctic, although nowhere is it very high. These various differences affect the distribution of vegetation but not the character of the zonal soils except in depth of thaw which is greater where warm summers prevail.

The presence of trees in the Subarctic and their absence in the Arctic appears to make little difference in soil profile development, particularly where permafrost is present. One of the main vegetative elements affecting soil development is the moss cover on the ground surface. In both the Arctic and the Subarctic, moss inhibits thawing in the warm season and retards soil profile development.

There is a great variety of parent materials throughout the permafrost region resulting in many types of soils, particularly mineral. The type of mineral soil depends largely on the geological nature of the parent deposits which may be glacial, alluvial, residual, colluvial, or aeolian in origin. The alluvial soils are the most suitable for agriculture, however, because they happen to predominate in the sheltered valleys where climatic conditions are most favourable for crop growth.

Relief affects the nature of the soil in several ways. It has some influence on the distribution of organic and mineral soils. Where moss growth is fairly rapid, as in the forests of the Mackenzie and Yukon River valleys, organic soils occupy the depressions except those filled with water. They also cover the lower slopes leading up from the depressions and much of the level land. In fact mineral soils are found only where surface drainage is good, as on river terraces. Slope aspect is a very important factor. Many steep north-facing slopes may be covered entirely with organic soils. In the Cordillera it is common to find organic soils with the permafrost table near the surface on north-facing slopes, while mineral soils without permafrost occupy south-facing slopes in the same locality.

None of the soils is very old and soil-forming processes are very slow. All but those in western Yukon Territory have originated since the Pleistocene glaciation. There are, however, considerable differences between alluvial soils of Recent age of the Mackenzie and Yukon Rivers and soils developed on alluvial materials that these rivers deposited when they were forming their present valleys.

5 Distribution of Soils

On the basis of broad differences in climate and vegetation, the soils of Canada's Arctic and Subarctic can be classified into zones. Each of these may be divided into subzones based on the nature of the underlying rock formations. Canada's permafrost region can be divided into the subarctic forested zone and the arctic tundra zone. There is a zone of transition but it is fairly restricted and generally the boundary is well defined.

Zonal soils of the forest and the tundra are of the same genetic type but generally differ in degree of development because of the milder climate of the subarctic zone. Other differences between the soils of the two zones are, however, important. In the arctic zone land surfaces covered with bare rock or with soils almost devoid of surface vegetation are considerably more extensive than in the subarctic zone, whereas the proportion of land covered with organic soils is much higher in the subarctic. The extensive Precambrian Shield areas lying north of the treeline around Hudson Bay and vast expanses of mineral soil with sparse vegetation in the Arctic Archipelago bear out this generalization. These areas are of little or no agricultural value because the climate is too cold.

Besides the division of permafrost soils into arctic and subarctic, there are also differences from one major physiographic region to another because of changes in parent material and relief. Five divisions can be identified. In the Precambrian Shield glacial ice moved over the hard rock landscape picking up what residual soil had existed previously, and deposited very little material so that the soil mantle is generally thin or absent. Mineral soils are generally coarse in texture consisting mostly of stony tills, gravels, and sands with some clays and silts. The soils are somewhat acid in reaction. Within the Shield area is the Hudson Bay Lowland bordering the west coast of Hudson Bay. The lowland lies near sea level having little relief and poor drainage. Its soils are primarily marine clays which are acid in reaction.

Areas of Palaeozoic limestone include the Arctic Archipelago and parts of the Mackenzie River Lowland. The relief varies from level plains to mountains in the Innuitian region of the Arctic Islands; and the soil mantle varies considerably in thickness but is generally thin, particularly in the high Arctic where soil profile development is greatly retarded. Because these soils are derived largely from the underlying rocks they are usually highly calcareous and strongly alkaline in reaction (Tedrow 1966).

The northern part of the Mackenzie River Lowland is comprised of Cretaceous shale. The relief varies from gently rolling to hilly and being typical of a young morainic surface is rather irregular. The soil mantle is usually fairly thick with few rock outcrops. Soil texture is fine grained generally, except where material from harder rock foundations has been brought in by glaciation. The reaction of the soils, derived generally from the soft underlying rocks, varies from weakly acid to alkaline depending on how much lime is present in the bedrock.

The Cordilleran Region has a complex pattern of relief and soil parent materials. Much of it was not glaciated. In the glaciated parts, most of the soils are coarse textured whereas the unglaciated parts have soils that are mostly medium textured and many of these are relatively free of stones.

The fifth soil division comprises recent alluvial soils, and is different from the previous four for several reasons. First, it consists of many separate widely scattered areas, and, second, each unit is too small to show on any but a large-scale map. They are of prime importance in northern agriculture, however, because they are situated in river valleys where access is easiest, and the climatic regime is such that agriculture in some form is possible. They are also usually well drained. Extensive areas of alluvial soils exist along Mackenzie and Yukon Rivers and their tributaries where certain garden and field crops can be grown. The soils vary in texture from fine sandy loam to silt loam, containing liberal amounts of organic matter. They are generally alkaline. Where the same are formed along rivers in Precambrian areas they are generally less extensive and of poorer quality. Because of the gentle and little-interrupted gradients of Mackenzie and Yukon Rivers in contrast to the tortuous courses of the rivers on the Shield, the two former rivers overflow their banks each spring during break-up, thus depositing new alluvium and adding to the alluvial soils developed there.

B   EFFECTS OF PERMAFROST

In Canada the generally accepted view is that permafrost exerts mostly harmful influences on agriculture in the North. In combination with the cool dry summers and thick moss cover, permafrost is recognized as a factor limiting soil profile development. Permafrost does not prevent crop growth but the low soil temperatures inhibit bacterial action and retard growth rates. Soils usually require drainage measures during the early years of cultivation, and land subsidence, due to the melting of ice in the soil, frequently imposes problems. On the other hand, progressive thawing of the frozen soil through the summer releases some moisture to partially offset the low rainfall. After a few years of cultivation the permafrost table recedes to a depth where it no longer limits the rooting zone of plants. However, low soil and air temperatures still limit the choice of crops.

1 *Depth of Thaw*

The depth of thaw is a critical factor in agriculture in permafrost regions because soil profile development, biological activity, and plant growth takes place in the thawed layer. The depth of thaw is determined by many factors, one of them being the vegetative cover (R. J. E. Brown 1966*a*). Leahey reported that in the Mackenzie River valley at Norman Wells, NWT, on 15 August, 1945, the mineral soil was thawed to a depth of 39 inches under 3 inches of organic cover in contrast to a thaw of only 20 inches under 6 inches of organic cover. Total destruction of the organic cover by cultivation resulted in a lowering of the permafrost table by 1 to 3 feet (Leahey 1954).

The locations of agricultural areas have a significant effect on the depth of thaw. In the Cordillera the thawed layer is thicker on south-facing slopes. In southern parts of Yukon Territory where permafrost is patchy, it is common to encounter permafrost on north-facing slopes but not on adjacent south-facing slopes. Russian investigators discovered greater depth of thaw on watersheds and less in meadows and river terraces (Tsyplenkin 1944).

Moisture and temperature conditions can be controlled to increase the depth of the thawed layer. Russian investigators accomplished this by improving the drainage of this layer and removing the organic cover and shade trees.

2 *Thermokarst*

Forms of micro-relief detrimental to agriculture are prevalent where large quantities of ice occur in the soil. When an area is in its natural undisturbed state before cultivation, the only thawing that occurs is in the top few inches or feet during the summer. The removal of trees and ground cover in forested areas and the ploughing under of grasses or sedges in open sites changes the thermal regime of the ground.

When cultivation is initiated the depth of thaw increases and, if the ice content is high, its melting results in severe differential settlement and subsidence of the land surface. Melting of a very large mass of ice may cause settling that results in the formation of a small pond. Many accounts of this thermokarst phenomenon appear in permafrost literature dealing with agriculture. In Alaska, Péwé described the problems encountered by farmers around Fairbanks. Large masses of ground ice exist in the silts of the Tanana River valley. When cultivated, the depth of thaw increased, ice in the perennially frozen soil melted and over a period of several seasons, subsidence and differential settlement resulted in the formation of mounds several feet high over so much area that whole fields had to be abandoned (Péwé 1954). Russian investigators have reported similar phenomena citing examples where small lakes have formed in thaw depressions in some cultivated districts (Tsyplenkin 1944).

3 *Thermal Regime*

Permafrost is detrimental to plant growth because the minimum temperature for plant growth is 42° F, and the temperature of permafrost is below 32° F. When the soil begins to thaw in the spring and the top few inches warm up above 32° F, conditions are still not suitable for plant growth as long as the temperature remains below 42° F. Even when the depth of thaw has reached its maximum, the lower part of this unfrozen layer is not capable of supporting plant growth where the temperature remains below 42° F. Generally, the soils are too cold at depths 6 to 18 inches below the surface although the depth of thaw may be several feet. This is particularly critical where the length of the growing season is very close to the length of time needed for plant maturation. It is essential, therefore, to warm the soil in the spring as quickly as possible by ploughing and other means. Once the temperature of the root zone rises above 42° F, growth will proceed. This situation prevails until the air temperature again falls below 42° F and inhibits plant growth at the surface. At the end of summer the temperature eventually falls below 32° F, and frost again penetrates downward from the surface.

   The degree of exposure to the sun influences the thermal regime of the soil. Soils on north-facing slopes differ from those of south-facing slopes in having shallower depths of thaw, higher moisture content, and lower surface soil temperature. The difference is such that north-facing slopes are frequently cold, wet, and peaty, with a cover of moss or low shrubs and scrubby trees in contrast to south-facing slopes which have relatively well-drained mineral soils under moderately dense stands of trees.

4 *Moisture Regime*

The moisture regime of cultivated plants is intimately affected by permafrost. As early as the nineteenth century Russian permafrost investigators recognized that permafrost exerted a beneficial influence in agricultural districts where precipitation, and particularly rainfall, was scanty (Tsyplenkin 1944). It has been recognized for some years that the permafrost table is impervious to water and tends to hold precipitation in the thawed layer, preventing subsurface drainage and producing swamp and peatland conditions. Thawing of the frozen ground during the summer provides water from melting ground ice. Rainfall is supplemented by a gradual release of frost-held moisture. If water within reach of the plant roots becomes depleted, a modification of dry farming methods could be employed.

Despite these statements, there is some disagreement on how beneficial permafrost really is in this regard. Gasser states that once the thawed layer has deepened
sufficiently for plant growth, the release of water from the thawing, perennially
frozen, soil is of very little consequence (Gasser 1948). This is particularly true
when agricultural land is first cleared and cropped. The subsoil is saturated completely with water because when the ground is covered by natural vegetation
thawing is limited to 2 feet or less. Under cultivation, the permafrost table recedes
to depths of 6 to 8 feet, especially in fields having a southern exposure. Free water
in the soil sinks as thawing proceeds, the crop plants use some of it, and there is
also loss by evaporation. The sum of these losses usually exceeds the summer's
average precipitation. The inevitable result is that subsoil on sloping land becomes
depleted of water. Once this has happened, crop plants depend exclusively on the
water that sinks in from the surface.

When the frost table lies at a shallow depth, the thawed layer is thin and is well
supplied with moisture. But this is the time when air and soil temperatures lie
between 32° F and 42° F. Even when ambient temperatures reach 42° F plant
activity is very low and moisture consumption by the plant is at a minimum. As
the depth of thaw increases, temperature conditions improve for plant growth but
the suprapermafrost moisture is also descending and the moisture content is restricted where root development is taking place in the upper part of the thawed
layer. In this way, the suprapermafrost water may be insufficient for plant requirements (Tsyplenkin 1944).

Russian investigators in Siberia discovered by field measurements that permafrost retards the upward movement of water into the thawed layer away from the
permafrost table, and at the same time contributes to the movement of capillary
water and water vapour from the upper to the lower part of the thawed layer.
Therefore, although permafrost is beneficial in so far as its thawing gradually
releases water to the active layer, yet it is detrimental because of its influence on
water movement in the thawed layer.

An "overdry" horizon develops in the thawed soil between the seasonal frost
and the permafrost. Capillary water is drawn upward to the seasonal frost line and
downward to the permafrost table. As a result the capillary connection between
the suprapermafrost water and the surface water may be broken. This is a particularly important phenomenon in the southern part of the permafrost region
where seasonal frost frequently does not penetrate to the permafrost table and a
thawed layer of soil persists between the two frozen layers. Plant roots in this
zone may have a water deficiency. It may be assumed therefore that capillary
water is greatly influenced by both permafrost and seasonal frost. This is particularly critical because it is the capillary water that contributes most of the water
required by plants (Tsyplenkin 1944).

Drainage of perennially frozen soils is often impeded by the imperviousness of the permafrost table. Korol reports that almost invariably virgin land needs some improvement, particularly drainage, to make cultivation possible. Light-textured soils such as light loam, sandy loam, and sandy soils that allow greater water infiltration should be reclaimed first. The permafrost table is generally deeper, and water and air penetration is easier than in heavier soils. Alluvial and Half Bog soils are usually poorly drained, although many of the latter need only a little artificial drainage to be suitable, at least for hay and pasture (Korol 1955).

Runoff and sheet erosion may be serious problems on long slopes exceeding 3.5 to 5 per cent with frozen subsoil where rainfall is fairly heavy, as on windward slopes of the Cordillera, or where snow accumulation in spring is heavy. Strip cropping is required on fairly steep slopes. Terraces produce mixed results. On shallow soils their construction buries too much surface soil. The subsoil under them is impermeable during spring thawing, but ice may block the channels and have the effect of concentrating water to initiate more serious gullying than would occur without terraces. Once gullies start, particularly in light-textured soils, they increase in size very rapidly. Special types of gullies are initiated by the melting of ice blocks in the perennially frozen soil. Although such gullies superficially resemble ordinary gullies, little or no soil material flows from them. If, however, the contour of the land is such that surface runoff does become concentrated in these gullies, they may deepen and become ordinary gullies.

Solifluction on slopes results in mass movement of soil downslope. Freezing and thawing of wet soils cause mechanical disintegration of the soil structure, changing it into a structureless hardened mass with a thick strong surface crust. The thawed layer increases in volume on freezing. Because of the unevenness of the penetration of the seasonal frost line, soil slurry is forced upwards in localities where the seasonal frost is thin, resulting in the formation of frost mounds which may grow to heights of several feet over a period of years. This is the opposite of the thermokarst phenomenon described previously, i.e., when differential ground settlement results in an uneven hummocky surface. In both cases, the cultivation of such land is difficult if not impossible (Korol 1955).

5 Nutritive Regime and Organic Content

Permafrost exerts a strong influence on the nutritive regime of cultivated plants because of the low temperatures imparted by it to the thawed layer. This means low temperatures in the root zone, lessened microbiological activity, and suppressed aerobic fixation of nitrogen and nitrification. In Siberia it was observed that the absorption of nitrogen by plants did not begin until the middle of July when the permafrost table had receded to a considerable depth, whereas crops and other plants were sown in mid-May. Therefore, in the first month of plant growth there is very little available nitrogen and the increase of vegetable matter is slow. Toward July the tempo of soil biological processes begins to increase rapidly and reaches a maximum in August when the period of the greatest nutritive needs of the plant has already passed.

The availability of potassium and phosphorus are also affected by low temperatures in the soil, their availability increasing with a rise in temperature. The same is true for minor nutrients such as manganese, cobalt, and molybdenum. Because of the low temperatures imparted by permafrost and the low bacteria count and activity, the decomposition of organic matter and the formation of humus is retarded. Leaching of nutrients into lower horizons of the active layer is retarded also (Tsyplenkin 1944).

In tundra soils the upper horizons are acid but lower horizons are usually slightly acid to slightly alkaline. Therefore, after clearing and ploughing, the soil becomes a mixture that is only slightly acid and well supplied with exchangeable calcium and magnesium. Ploughing must be fairly deep, however, to provide this situation where base exchange is possible (Kellogg and Nygard 1951).

In virgin conditions most potentially arable soils have a relatively high content of organic material in the "A" horizon. Even in lower horizons, there may be significant amounts of organic material in the form of buried leaves, twigs, roots, and logs. Once it is lost by burning, clearing, and erosion, it is difficult to restore because of the short season for green manuring of crops. Therefore, in Alaska for example, animal manures, peat, and compost are used generally. In northern regions, the natural organic material decomposes slowly but once it is gone it is replaced very slowly. Russian soil scientists report good success in the Subarctic with lupines for green manure, in rotation with potatoes. On dairy farms, barnyard manure has been used alone or composted with peat and other organic matter. Most peats are too raw for direct use but some that are decomposed are satisfactory (Kellogg and Nygard 1951).

The inherent fertility of virgin soils in permafrost regions is very low – after only three to five years of cultivation, agricultural land requires heavy fertilization to support crop growth. The immaturity and retarding of soil profile development because of low temperatures is responsible for this. Much labour is wasted if fertilizer is not used – commercial varieties containing nitrogen, potassium, and phosphorus, as well as manures. Nitrogen deficiency is the most critical problem followed by phosphorus. Manure must be partially decomposed because of the relatively feeble biochemical activity in the soil and best results are obtained from a mixture of manure and commercial fertilizers. The soils that are cultivated in the permafrost region are mostly acid soils that require lime also. The best crop yields generally come in the second year after virgin land is ploughed, after which yields fall off rapidly and may even fail to seed. Thereafter, large amounts of organic manure and mineral fertilizers are needed to maintain even a medium output. Despite their higher potential fertility (compared to mineral soils), peaty and peat-glei soils are more difficult to cultivate because of the lesser depth of the thawed layer (one to two feet). Therefore, these soils must be drained and the moss and peat burned. As the permafrost table recedes the land can be used for crops.

Carbonic acid in the soil water is an active agent in breaking down, dissolving, and carrying nutrients out of reach of plant roots. The lower the temperature of the soil the greater the quantity of carbon dioxide dissolved in the soil solution and, thus, the higher the carbonic acid content. The amount of air dissolved in water is 50 per cent higher at $32°$ F than at $68°$ F, making the carbonic acid content correspondingly higher. Proximity of the permafrost table to the ground surface and its low temperatures is detrimental in this respect (Tsyplenkin 1944).

Permafrost prevents the spreading of plant roots to deeper horizons and the low soil temperatures reduce the rate of assimilation of moisture and nutrients. A considerable quantity of nutrients is required in the spring and plants with shallow rooting systems which can absorb the moisture and nutritive substances are preferable (Tsyplenkin 1944).

6 Cultivation Techniques

Techniques which have been used in Alaska are described by Kellogg and Nygard (1951). The land is usually cleared with bulldozers – a task which must be done with care in order to avoid removing the organic surface layers. Because freshly cut trees and moist stumps are difficult to burn, because of the high moisture content of the soils, these are pushed into windrows at the edges of the fields and burned later. Partially burned remains are concentrated into piles with the bulldozer and burned again until they are finally destroyed. The upper horizons may be destroyed if the bulldozer blade is set too deeply so that one method used to save the organic "A" horizon involves bulldozing in winter with the blade set high to shear the trees from the frozen soil. The blade may then be used later to remove stumps and large roots.

Individual fields are generally quite small and therefore tools and machinery such as tractors may be smaller than in temperate areas. Although the long summer days compensate somewhat for the low rainfall, the short growing season and the perennially frozen condition of the soil, it is important that the surface soil warms up as quickly as possible in the spring: deep ploughing helps here. A delay of only a few days may spoil the chance of a good yield, particularly because of the short growing season, therefore, mulches and cover crops should be avoided – despite the advantage of building up the organic material, controlling erosion, and reducing the blowing of soil exposed to dry winter winds.

Because of light rainfall, irrigation is helpful. Excessive thawing of the soil must be avoided, however, because of the danger of waterlogging soils, especially those underlain by permafrost. For this reason, and because fields are generally small, widely separated sprinkler irrigation seems to hold promise as a practical method.

7 Types of Cultivated Crops

Types of crops that can be grown in permafrost areas are limited. On the tundra soils of Alaska, there are only a few small parcels of cultivated land on which hardy vegetables such as radishes, turnips, and cabbages are grown, and here large amounts of organic matter are added to build up the soil.

On subarctic brown forest soils an ideal crop is a perennial legume that resists winter killing – a good seed crop like Siberian alfalfa is difficult to grow because of its low resistance to winter killing. Relatively good summer pasture is available but oats and peas produce poor yields for hay and silage, and hay is difficult to cure naturally because of the low summer rainfall. Weeds create a serious problem becase they thrive in the long summer days. Therefore, it is difficult to devise a good cropping system that will maintain the organic material, protect the soil, and furnish enough roughage for winter feed (Kellogg and Nygard 1951).

Potatoes and other crops have been grown in the Siberian tundra as far north as the Taymyr Peninsula. Barley, oats, spring wheat, spring rye, and other cereals thrive in the coniferous forest belt. The variety of crops increases southward in the deciduous–coniferous belt with the addition of buckwheat, millet, peas, and others (Korol 1955).

At best, the selection of crops is limited. The short growing season combined with the detrimental effects of permafrost make cultivation a gamble.

C AGRICULTURE IN NORTHERN CANADA

Agriculture is limited almost exclusively to the valleys of the Yukon and Macken-zie River systems. Here about five million acres of suitable soils for crops may be found although very little of this is being cultivated at present. Papers by Nowosad (1959 and 1963) give a general survey of the current limited agricultural activities in northern Canada. These accounts provide the main source material, with supple-mentary information from other references.

1 Mackenzie River Valley

The Valley or Lowland can be divided into three zones. The first is the Northern Plains Zone extending south of a line from the south side of Slave River delta to Camsell Bend. Its mean annual air temperature is above 25° F and permafrost is patchy. Second is the Subarctic Zone which lies north of the first and extends to about 68° N. The mean annual air temperature varies from about 18° F to 25° F and permafrost is discontinuous to continuous. Third is the Arctic Zone which lies north of the Subarctic Zone. There mean annual air temperature is about 15° F or less, and permafrost is continuous (Dawson 1947).

The soils of the Mackenzie River lowland can generally be divided into upland mature soils and lowland recent soils. There are, however, extensive tracts of undeveloped or poorly drained soils where heavy applications of fertilizers are needed for continual cultivation. These factors combined with the low temperatures, low rainfall, and permafrost, restrict land-use possibilities even in the lowlands. Permafrost impedes drainage and so water collects in low places to form lakes of all sizes and shapes which then spill from one to another forming a very indefinite drainage pattern. Thus, the percentage of arable land is very small and is confined mostly to the vicinity of the larger streams. Away from these water courses, frozen swamps, peat, and shallow lakes cover a large percentage of the surface (Albright 1933). Depending on latitude, there is a thaw of 3 to 10 feet in cultivated soil, 2 to 3 feet in wooded soil, and less in peat (Robinson 1945b, Albright 1933a).

a / Present agriculture

Hay River is situated on the south shore of Great Slave Lake in the southern part of the Northern Plains Zone. Here permafrost occurs in scattered islands, restricted to areas having a thick peat cover. Cultivation of these areas has caused the permafrost to disappear and a variety of garden crops has been produced successfully for a number of years.

North of Great Slave Lake, the next settlement on Mackenzie River is Fort Providence where, in the settled area, permafrost has not been encountered. Wheat and barley have been grown here but present cultivation is restricted to garden vegetables (Chambers 1907).

Fort Simpson, situated on an island at the junction of Liard and Mackenzie Rivers, near the northern limit of the Northern Plains Zone, was founded in the early years of the nineteenth century. Since the end of that century, various sections of the island have been cleared and cultivated. Where permafrost exists, there is some correlation between the annual depth of thaw and the number of years since an area was cleared. The depth of thawed soil varies from a few feet, in areas cleared only a few years ago, to tens of feet, in areas cleared several decades ago (Pihlainen 1958). In 1946 the federal Department of Agriculture established an experimental farm to conduct an active programme of research on soils and farming. Since then, spring wheat has yielded forty bushels per acre and coarse grains have ripened every year. Root vegetables, tomatoes, berries and tree fruits have been grown every year.

Permafrost is more widespread and nearer the ground surface in the Subarctic Zone than in the Northern Plains Zone. Although precipitation is scanty, evaporation is low and some moisture is released to the root zone by thawing of the frozen ground. Cultivation in the settlements is restricted to vegetable gardens because of the short growing season. At Norman Wells the settled area was stripped of the natural vegetation in the 1940s causing the permafrost table to recede to such a great depth that it has no effect on garden vegetables. Ninety miles north at Fort Good Hope the soil thaws to a depth of four feet or more in cultivated plots. Cabbages and onions are grown every year and potatoes, turnips, and barley have ripened occasionally (Figure 40) (Albright 1933*b*). Vegetables are grown at Thunder River and Arctic Red River in garden plots where the permafrost has receded to depths of three to four feet after several years of cultivation (Robinson 1945*b*).

Aklavik is situated in the Arctic Zone, within the continuous permafrost zone. The Roman Catholic Mission was established there in 1926 and in the following year the ground was broken for a garden. The depth to permafrost before cultivation was only about one foot and radishes, lettuce, and potatoes were planted. Of these the radishes thrived but the lettuce was small and the potatoes grew only six to eight inches tall in the first year. Cabbages were also grown successfully. By 1930 the annual depth of thaw had increased to two feet. It was discovered that generally only short-rooted and shallow-rooted plants could be grown successfully (Hutton 1946). Seeds placed too deep in the ground germinated too late and often turned what should have been a fairly early crop into a late summer one in danger of being killed before maturity by autumn frosts.

At Inuvik the development of a small experimental garden by the federal Department of Agriculture has provided information on the recession of permafrost and the improvement in crop production with cultivation. The garden is on a gentle slope with a southwest exposure and formerly was covered by trees and about 12 inches of moss and peat. In June 1956, after removal of the trees and about 10 inches of the organic cover, frozen soil was encountered at a depth of 2 to 3 inches. By 29 August the ground had thawed to 24 inches. Cabbages, carrots, potatoes, and turnips were planted but germinated poorly and grew slowly.

In 1957 the same vegetables were planted. By 1 August the soil had thawed to three feet and in the early fall to four feet. Fair production of some garden crops was obtained, with potatoes, for example, yielding 69 bushels on an acre basis. By the fall of 1958 the ground had thawed to 58 inches; most vegetables grew very well and potatoes yielded 185 bushels on an acre basis. In 1959 crop production equalled or exceeded the 1958 yield and the ground had thawed to a depth of 70 inches by the fall. In 1962 the permafrost table was at a depth of six feet.

At Tuktoyaktuk, on the arctic coast northeast of the Mackenzie delta, one or two small garden plots have produced a few vegetables. This settlement is in the tundra and the vegetative period is very short. Under the present climatic conditions it is not possible to grow plants on anything but a very limited basis.

An interesting experiment was conducted about 1967 by McMaster University at Resolute, NWT in the Arctic Islands in the northern part of the continuous zone. The active layer is only about one foot thick and growing conditions are extremely severe. Ground plots were laid out and covered with various types and amounts of dark-coloured material such as soot and ashes. Thawing of the frozen ground was accelerated and soil temperatures increased sufficiently beneath the darkened surfaces to permit the growth of a few potatoes and cabbages. The conditions are so severe that growth is extremely limited and only possible on an experimental basis.

b / Potential developments

In 1907 a Canadian parliamentary committee saw little hope that the Mackenzie River lowland could ever support a purely agricultural community or that its products could ever compete on the world market. It seemed likely that agriculture would remain a subsidiary activity, its development depending upon the establishment of a local market. It was recognized that the alluvial lands along Slave River, the upper part of Mackenzie River, and the country bordering Liard River for some distance above and below Fort Liard appeared to hold most promise (Chambers 1914).

FIGURE 40 Cultivated plot on permafrost at Fort Good Hope, NWT, in northern part of discontinuous permafrost zone on a terrace of the Mackenzie River.

In 1955, almost 50 years later, the Commissioner of the Northwest Territories submitted a brief to the Gordon Commission on Canada's Economic Prospects. He reported that surveys to date indicated that there may be between 1,000,000 and 1,500,000 acres of arable land in the Northwest Territories. These included some 500,000 acres of good ranch land in the Slave River basin and between 100,000 and 200,000 acres of mixed farm land in the Liard Valley. There were various sections of land which would be suitable for vegetable farming along Hay River and also the length of the Mackenzie River. He concluded that it was unlikely that land in the Northwest Territories would be used to produce agricultural products for shipment to other parts of Canada within the next few decades because there was probably sufficient land closer to markets with larger populations. With population growth on the Mackenzie River and in the vicinity of Great Slave Lake, however, these arable lands could be expected to develop for local markets (Robertson 1955b).

The Experimental Farm Service of the federal Department of Agriculture has conducted a number of surveys in various parts of northern Canada to assess the agricultural potential based on surveys of the soils and native vegetation. The areas which have been surveyed in the Northwest Territories are the Hay River area, the Slave River lowland, Mackenzie River downstream to Fort Simpson, Fort Nelson River, and Liard River.

In the Hay River area the mean annual air temperature is 24° F. Although total precipitation is low, much of the land is poorly drained. Soil temperatures in summer are low even in well-drained mineral soils. Permafrost is patchy and is found mostly in peat bogs which are unsuitable for agriculture because of the difficulty of establishing and maintaining drainage. The recent alluvial soils consist of silt or silty clay-loam in the upper horizons and sand in the lower horizons. The depth of the sand is variable, ranging from one to several feet, with drainage through the soil varying from excessive to poor. In the poorly drained depressions surface peat is 6 to 18 inches thick and the permafrost table lies 18 inches below the ground surface. The most promising type of agriculture appears to be vegetable gardening in areas where permafrost is absent or too deep to affect the thermal regime of the root zone, and also where the surface peat is thin (Leahey 1953).

The Slave River lowland is bounded on the east by the Precambrian Shield and on the west by the Palaeozoic Upland. Permafrost occurs in scattered islands in several types of soil encountered here. Whereas the soils with no permafrost have considerable agricultural potential, soils with permafrost would require some effort to bring them into cultivation. In one type of soil it would be possible to garden only if intensive management practices were employed, for example, this soil could be built up by incorporating at least part of the peaty surface soil into the mineral soil and by the use of adequate amounts of fertilizers. It would appear that in most seasons irrigation is a necessity for crop production. In another type of soil difficulties could arise from the ground subsidence and from the development of irregular relief due to melting of the large quantities of ground ice which are known to exist. The agricultural possibilities of the organic soils are not good because of their high acidity, surface micro-relief, poor drainage and shallow depth of the permafrost table (Day and Leahey 1957).

Along Liard River permafrost is not found in any areas suitable for agriculture and therefore should not be a problem in future cultivation (Leahey 1944).

Along Mackenzie River the best areas for further development are near the banks of the river – cultivation away from the river does not appear feasible because of severe climatic conditions.

2 *Yukon Territory*

The mountainous terrain of Yukon Territory limits agriculture to glacio-lacustrine deposits in a few valleys. The climate is dry and nowhere is the soil weathered deeply. Most soils are forested but there are a few indications of a definite leached layer so characteristic of wooded soils in northern latitudes. In a few local areas there is some incipient development of the characteristic light grey A_2 horizon of podzolic soils under spruce on light-textured soils. Mineral soils are comparatively low in organic material (Dawson 1947).

The Yukon Territory lies mostly in the discontinuous zone, permafrost being patchy in the south and widespread in the north. Local variations are caused by the mountainous terrain: permafrost is thicker and the depth of thaw is less on north-facing slopes than on adjacent south-facing slopes. On the Alaska Highway unfrozen soils predominate between Teslin and Kluane Lakes; west of Kluane Lake perennially frozen soils predominate, even on south-facing slopes; north of Carmacks perennially frozen soils are increasingly widespread and are found at shallow depth on south-facing slopes in the Dawson and Mayo areas, unless cleared by man or fire (Leahey 1943).

Along Stewart River, between Mayo and Minto Bridge in central Yukon Territory, the valley slopes are covered with spruce and moss under which permafrost is found at a depth of about one foot. Along the streams in the Mayo area there are small scattered areas of unfrozen silt. Downstream, halfway to Stewart Crossing, the river flows through a broad valley and the forested south-facing slopes are frozen at shallow depth. Towards Stewart Crossing there are a few river flats, on one of which is located the Maisy Mae Ranch – the only farm on the Stewart River. Generally these river flats have a dense spruce and thick moss cover and permafrost is found near the river. One exception is the Yukon River flats, which are forested but are not frozen as near the river (Leahey 1943).

a / Development of agriculture
Agriculture in Yukon Territory began during the Klondike gold rush. During the early years of the twentieth century there was some farming, gardening, and horse ranching around Dawson and at scattered points along the water route from the south. The population, supported by the mining industry of the Klondike, exceeded 20,000 and some sizable farms developed. No statistics for agriculture in Yukon Territory exist for the years prior to 1931, but it is known that heavy importations of food continued throughout the boom period. The steady decline in population from 27,219 in 1901 to 4,157 in 1921 had, however, an adverse effect on agricultural development (Dawson 1947).

In 1917 the federal Department of Agriculture began a limited amount of experimental work at Swede Creek 8 miles west of Dawson. It was discovered that yields of 200 bushels of potatoes per acre could be obtained if the land was manured and heavily fertilized. On nearby unmanured land the yield was only 68 bushels per acre because of the inherently low fertility of permafrost soils. Experimental work was discontinued here in 1925 (Dawson 1947).

Trials with roots, cabbages, cauliflowers, peas, beans, potatoes, barley, oats, and fodder corn were conducted from 1932 to 1934 at Carmacks. Grasses, clovers, and cereals were grown from 1936 to 1938 at Carcross where summer frosts are a hazard (Dawson 1947). In 1941 there were 26 farms in Yukon Territory with a total area of 2,781 acres (Robinson 1945a). In the late 1950s potatoes were produced commercially at Carmacks and Keno Hill. Greenhouses and hot-beds were being used to produce tomatoes and cucumbers for the local inhabitants. Potatoes, carrots, and most common vegetables could be grown at Dawson even farther north.

In 1944 the federal Department of Agriculture established an experimental sub-station at Mile 1019 on the Alaska Highway, 106 miles west of Whitehorse, which is still operating. Its objectives were twofold: one aim was to determine the agricultural possibilities of the arable land in the Takhini-Dezadeash Valley in which it is located; the second was to serve Yukon Territory in determining the practicability and location of agricultural experiments in the territory, assuming some increase in population and market possibilities (Leahey 1943).

b / Potential developments
The largest block of arable land in Yukon Territory is an area in the Takhini-Dezadeash Valley covering approximately 120,000 acres. South of the Alaska Highway near the Yukon–British Columbia border are the Tagish and Little Atlin Flats where arable land covers about 8,000 acres (Dawson 1947).

In central Yukon Territory there are about 6,000 acres of arable land on the upland soils around Dawson and in the Mayo area there are parcels of arable

land on Stewart River (Dawson 1947). These areas are valuable in that they provide places where gardening may be successful, but it is doubtful if any of them
are large enough to permit any but very small farms. It is reported, however, that
these areas could supply the town of Mayo with most of the garden vegetables that
it requires (Robinson 1945*a*).

3 *Provinces*

Agriculture has been carried out extensively in the Peace River region of northern
Alberta and British Columbia for many years. The federal Department of Agriculture maintains an experimental farm at Fort Vermilion, Alberta where many
types of crops are grown successfully. Although it is situated in the discontinuous
zone, no permafrost exists at the farm now. Scattered islands of permafrost occur
in peat bogs in the Peace River region but they have had no effect on agriculture.

Permafrost is widespread, however, at Fort Chimo near Ungava Bay in northern
Quebec in the northern part of the discontinuous zone where the federal Department of Agriculture established an experimental station in the mid 1950s. The
possibilities of agricultural development at this subarctic location were investigated
in 1954 when a good crop of potatoes was observed to be growing in a plot where
the permafrost was at a depth of forty-seven inches. Various types of vegetables
including cabbages, cauliflowers, lettuce, beets, and onions have been grown
successfully. Experiments in raising sheep and poultry have also been carried out
with promising results.

One procedure was to work the garden with a small tractor and fertilize it after
the soil had thawed and dried sufficiently on the surface. Polyethylene strips three
to four inches wide were laid out to serve as a mulch. Soil temperatures
increased and thawing progressed more rapidly beneath the plastic than in exposed
areas. Slits were cut in the mulch and transplants were put in place.

D   FORESTRY

The utilization of forest resources through harvesting trees for timber or pulp
can be considered as belonging to agriculture. Very limited use of these resources
has been made in the permafrost region of Canada except in a few sawmilling
operations on the Peace River and at a few points along the Mackenzie River.
Some of the timber piles used at Inuvik for building foundations were obtained on
the Peel River, a northern tributary of the Mackenzie River. Little work has been
done on this subject but an account by the federal Department of Forestry of the
present situation is reported here (Robinson 1965).

Virtually all of the forested part of the permafrost region is confined to the discontinuous zone which is occupied by the northern fringe of the boreal forest. This forest region extends into the continuous zone in the lower Mackenzie River valley and Mackenzie delta. Knowledge of how permafrost develops and how trees react to frozen ground may provide information essential to management of boreal region forests. It appears that under the climatic conditions of the northern boreal forest an acid litter develops under the trees followed by growth of a moss cover. The moss fills with snow and ice during the early snowstorms of the autumn and thus serves as a conductor of heat from the ground during the winter. In the summer the surface of the moss dries and acts as an insulator hindering the transfer of heat to the underlying mineral soil. As a result only a portion of the winter frost may be dissipated during the summer and permafrost will develop.

The exact influence of the moss cover is not fully understood, although research is being conducted in this field by the Pulp and Paper Research Institute of Canada, the federal Department of Forestry and other agencies. The first obvious effect is a lowering of soil temperatures which will discourage all trees or plants whose roots cannot grow in a cold soil. The acid litter which develops, especially under a spruce forest, is low in available nutrients, particularly nitrogen, so that few of the nutrients can be assimilated, regardless of the potential richness of the soil. The litter tends to make the underlying soil surface impervious to moisture and an artificial water table develops which prevents soil aeration. This condition is also found in sandy soils where impervious "hard pans" can develop with the same ultimate results. Exact information on the growth-inhibiting effect of this acid litter seems to be unavailable so it is not certain whether the reduction of soil temperature or the increase in acidity of the active soil layer is the growth deterrent. Probably both are contributing factors.

If the development of the acid litter and moss cover is prevented, spruce trees will grow quite satisfactorily in a very shallow active layer in permafrost regions. At Fort McPherson, NWT, the average depth of thaw at the end of July is only twelve inches. Steep slopes which have good drainage plus moisture seeping from the hillside develop moss layers very slowly and have good tree growth.

Forest development suitable for exploitation requires an active layer which will permit tree growth but the cost of preparing the soil must be economical. Early studies in Alaska indicated that the active layer had to be several feet thick but the Mackenzie delta forests show that trees can grow on a very thin active layer as long as the soil surface is warm. Thus the forester must keep the soil thawed and his major problem is to develop a method at a price which is not beyond the value of the final crop one hundred years later.

Trees grow very slowly on permafrost and the merchantable stands of the Mackenzie River delta are mostly about 200 years old although it is possible that timber of small sawlog size might be grown on favourable sites in as little as 100 years. These stands of timber are found, however, where periodic floods from ice jams cover the forest litter with rich silts. The floods incorporate the leafy matter into the soil and effectively kill the moss layer. Above the level of periodic flooding the trees are almost entirely scrub; for example, black spruce more than 300 years old only six feet tall have been observed.

Only a limited area is subject to such periodic flooding and more widely applicable procedures are required. Stripping the surface vegetation to accelerate thawing of the permafrost such as is done for agriculture is a very expensive process far too costly for the development of forest land, and there is no guarantee that a site will not degenerate quickly after trees begin to shade the soil.

On a hillside above Norman Wells, NWT, a serious fire in 1959 burned the forest and ground cover down to the mineral soil. Downhill below the burn, black spruce scrub and thick Sphagnum grew on the permafrost table which was only eighteen inches below the ground surface. Similar conditions were found uphill from the burn. On the burn area itself, however, the permafrost receded below the four-foot depth. Fire is used in the Scandinavian countries to release chemicals at the surface of moss layers and thus promote tree growth. It might be effective in northern Canada, especially if flat or gently sloping burn areas were ditched to provide adequate drainage. Considerable work is being done in southern Canada on the use of controlled burning and this may be used effectively in permafrost regions.

Uncontrolled burning is usually seriously destructive. Thin-soiled areas are normally the driest and with the litter removed by fire they degenerate to rock barrens. Swamp areas are usually too wet to burn and remain swamps. Since most of the bench land of the Northwest Territories is now so wet that fires would not affect it, burning, controlled or uncontrolled, would be difficult to utilize without extensive preliminary work.

In conclusion the conditions which lead to permafrost development also lead to forest site degeneration, even in areas where soil and summer air temperatures are suitable for tree growth. The forester's problem in such areas seems to be one of developing a "litter" layer which will permit warming of the upper soil horizons, but preventing this layer from acting as an insulating blanket which will foster permafrost.

E   CONCLUSION

Agriculture in Canada's permafrost region is only marginal and promises to be so for the foreseeable future. It is confined generally to vegetable gardening at settlements along water routes such as the Mackenzie and Yukon River systems and land routes such as the Alaska Highway. In addition, the federal Department of Agriculture has been conducting experiments in agriculture at several scattered localities. There are tracts of arable land which could be cultivated to provide produce for the local populations. These are at present small and scattered and so it is more convenient to import from temperate regions. However, these potentially productive areas, most of which lie in southern fringes of the permafrost region, could be developed when the need arises.

Apart from the small local demand for produce, agriculture in Canada's permafrost region is beset by a host of physical handicaps. Extensive glaciation has left large areas particularly in the Precambrian Shield with little (or no) soil, usually of poor quality. Low soil temperatures inhibit root development and lower the availability of plant nutrients because they hamper the moisture regime of the soil even although the thawing of frozen soil which contains ice releases water to plants in these regions of low rainfall. Melting of ground ice often creates such uneven micro-relief that ploughing and cultivation are impossible. Drainage may be hampered by impervious frozen subsoil and the accumulation and retarded decomposition of organic material results in high acidity requiring excessive lime application.

Perennially frozen subsoil is a product of the severe climatic conditions which handicap plant growth. The short frost-free period and the even shorter period when air temperatures are above 42° F – the minimum temperature for plant growth – is the main problem. The scanty rainfall over much of the North already has been mentioned. Another feature is the continuous daylight in summer which compensates to some extent for the shortness of the growing season. On the other hand, plants native to temperate climates seem to prefer alternating day and night because continuous daylight stimulates growth of the leafy parts of the plant at the expense of the fruit.

In the present discussion, considerable reference has been made to agriculture in Alaska and the USSR where there are problems created by the perennially frozen subsoil. In contrast, little has been written about such problems in Canada's permafrost region so that many of the examples that illustrate the effects of permafrost on cultivated plants have had to come from this foreign literature. Larger populations are dependent on agriculture in these regions of Alaska and, particularly the USSR, and permafrost has presented a more immediate problem than in Canada. There is no doubt that if agriculture develops in Canada's permafrost region, further examples of its effect on this phase of human activity will come to our attention.

9 Conclusion

An examination of the ways in which permafrost affects the various spheres of man's existence in northern Canada shows that it exerts considerable influence on his activities there. Whether the permafrost occurs in scattered islands as near the southern limit of the discontinuous zone, or whether it is widespread and hundreds of feet thick as in the continuous zone, it is a factor meriting serious consideration in the development of an area.

It is possible to divide the permafrost region into subregions of varying types of permafrost problems and the means of coping with them. For example, one well-established division is discontinuous and continuous zones. Areas of thawed ground are found in the former zone in contrast to the latter zone where permafrost is found everywhere beneath the ground surface. Therefore, it is sometimes possible to avoid permafrost in the discontinuous zone and employ methods used in temperate regions whereas it is impossible to avoid permafrost in the continuous zone. Construction and other activities in the discontinuous zone are complicated, however, by the erratic and often unpredictable occurrence of permafrost and the proximity of its temperature to 32° F.

In the construction of buildings, for example, permafrost may be avoided in the discontinuous zone and the structure designed for the properties of the soil in the thawed rather than the frozen state. Frozen ground can often be thawed and prevented from reforming by stripping the vegetation and allowing the heat from the building to maintain the soil in the thawed state. Problems can arise, however, where the permafrost, which is close to 32° F, may thaw slowly during the life span of the structure. In the continuous zone, it is impossible to avoid permafrost and buildings must be designed in most cases for the properties of the soil in the frozen state.

Roads and railroads are slightly different because they extend many miles over which the permafrost may change in extent and thickness. In the discontinuous zone, it is possible to avoid permafrost in some areas and design for the properties of the soil in its thawed state. In the continuous zone, the soil is perennially frozen everywhere and construction techniques must be modified accordingly.

Mining presents a somewhat different situation because the site of this activity is determined by the location of the orebodies. In the discontinuous zone, an ore-body may be either unfrozen, partly frozen, or completely frozen depending on the extent and thickness of permafrost in the vicinity. In the continuous zone, an orebody may lie in or beneath the permafrost depending on the thickness of the frozen ground.

The relation of agricultural activities to permafrost distribution is similar to that of buildings. In the discontinuous zone, it is frequently possible to avoid permafrost whereas in the continuous zone permafrost is encountered everywhere near the ground surface.

Although there are broad differences between the discontinuous and continuous zones, it is difficult to subdivide these zones. This is because human activities are concentrated at scattered points separated by vast tracts of uninhabited land and the development at any one spot is conditioned by local factors. The method employed at any point of activity is to evaluate the existing natural conditions and use the best possible techniques of construction and exploitation keeping in mind transportation costs, availability of local materials, labour and other cost factors. It will probably be many years before human activities are sufficiently widespread across the permafrost region to enable the subdivision of each permafrost zone into areas based on categories of permafrost problems and their solutions.

Because permafrost occurs widely in Alaska and the USSR some details of their experience have been included. The physical properties and effects of permafrost are similar in the three countries and the methods of coping are generally similar. In each country it is recognized that kept frozen, soils having high ice contents provide good bearing capacity for structures, but that allowed to thaw, they lose this bearing strength. Structures founded on such soils will be damaged or even destroyed. The location of transportation routes in permafrost regions is governed by the same principles in each country: to locate the route on well-drained soils with low ice contents and to prevent thawing of the base course – which causes severe settlement of the grade and makes the route unusable. Mines in permafrost regions of the three countries face such problems as water seepage into shafts and drifts resulting in massive ice accumulation, and frozen ore which resists blasting and other conventional extraction methods used in temperate areas. Agriculture is hampered by such problems as the presence (because of the proximity of the permafrost table) of bodies of ice in the ground; these thaw when the land is cleared, causing such large depressions that farm machinery cannot be used for cultivation. The above are but a few of the major problems which are an indication of the many-sided influence of permafrost.

Economic factors and the development of certain practices mean that these countries do differ in their approach to the problem of permafrost. For example, in some settlements in Canada and Alaska, water and sewage lines are carried in utilidors above ground. In the USSR, on the other hand, where utilidors are rarely used, one method is to excavate a trench several feet deep, bury the pipe in this trench, backfill with coarse-grained soils which are not frost-susceptible, and heat the water or sewage to a sufficiently high temperature to prevent freezing. Another method is to place the pipes in a precast concrete conduit which is half buried in the ground above the permafrost table. In Canada buildings are mostly one or two storeys high and frequently of wood frame construction on wood pile foundations, in contrast to the USSR where many buildings up to ten storeys are constructed of concrete blocks or precast concrete panels on precast reinforced concrete piles. The difference here appears to be that in the USSR large cities with large populations have arisen in the permafrost region as a result of widespread exploitation of natural resources; and the need for large structures to accommodate the various functions of these cities has been recognized even though the cost of construction of such buildings on permafrost is high and greater than in temperate regions. On the other hand, settlements in Canada's permafrost region are small and few in number. The need for large multi-storey buildings is not pressing and smaller wood frame buildings are adequate and perform well. Thus although it would be possible to construct large masonry buildings in Canada's North they are not required.

Permafrost is only one of many factors which hamper developments in northern Canada. The severe climate and brief summer season drastically shorten the length of the construction and agriculture seasons. Long winters raise annual heating costs. The long period of continual darkness in winter is both a physical and psychological deterrent to any operation. Land transportation routes into the North are limited and water routes are usable for only a few months in the summer. Because of the remoteness of the area from southern regions and the difficulty of marketing exploited resources so as to compete economically with similar ones in temperate regions, resource exploration and exploitation is limited. It is not lack of technical knowledge so much as logistical problems and economic considerations which have caused the development of the permafrost region of Canada to be slow.

Permafrost is only one of a number of factors influencing activities but it is a phenomenon common to the entire north of Canada. Its generally adverse effects result in increased costs of construction and operation thus impeding the possibilities for settlement and the prospects of transportation, mining and agriculture.

In spite of these special problems not encountered in temperate regions, development of the permafrost region will continue at an accelerated rate – the ability to cope has already been demonstrated, and improvements in technology, combined with careful site selection and evaluation of the permafrost problem, indicate that it is possible for man to live in the North. The establishment of new roads for access to resources, more surveys of the mineral wealth of the area, and examination of the agricultural potential of perennially frozen soils all suggest that, with the pressure of world population growing as it is, permafrost and other adverse natural conditions will not prevent increasing human activities in this region, which is one of the few large empty areas still remaining on this planet.

References

AHO, A. E. (1966). "Exploration Methods in Yukon with Special Reference to Anvil District," *Mining J.*, vol. 87, no. 7, pp. 50–8.

ALBRIGHT, W. D. (1933*a*). "Gardens of the Mackenzie," *Geogr. Rev.*, vol. 23, pp. 1–22.

—— (1933*b*). "Crop Growth in High Latitude," *Geogr. Rev.*, vol. 23, pp. 608–20.

ALTER, A. J. (1950). "Relationship of Permafrost to Environmental Sanitation," presented at the Alaskan Sci. Conf., 9–11 Nov., Washington, DC, 11 pp. (mimeo.).

ANDREWS, J. T. (1961). "Permafrost in Southern Labrador-Ungava," *Can. Geogr.*, vol. 5, no. 3, Autumn, pp. 34–5.

ANNERSTEN, L. J. (1964). "Investigations of Permafrost in the Vicinity of Knob Lake, 1961–62," in *Permafrost Studies in Central Labrador-Ungava*, J. B. Bird, ed., McGill Subarctic Res. Papers no. 16, pp. 51–137.

ARMSTRONG, J. (1961). "The Problems behind Arctic Drilling," *Oil in Can.*, 9 Nov., pp. 35–40.

Atomes (Paris). (1954). "Siberian Oil Deposits," Jan., p. 30.

BABKOV, V. F. (1954). "Dorogi v Vechnomerzlykh Gruntakh" (Roads on Permafrost), in *Avtomobil'nye Dorogi* (Automobile Roads), V. F. Babkov, Nauchno-Tekhni-cheskoe Izdatel'stvo Avtotransportnoy Literatury (Science-Technical Publ. of Automobile Transport Literature), Moscow, pp. 119–21 (Text in Russian), abstract (SIP 12719).

BARNES, D. F. (1966). "Geophysical Methods for Delineating Permafrost," *Proc. Permafrost International Conf.*, Nat. Acad. of Sci., Nat. Res. Council Publ. no. 1287, pp. 349–55.

BELCHER, D. J. (1948). "Determination of Soil Conditions by Aerial Photographic Analysis," *Proc. Second* International Conf. Soil Mechs. Foundation Eng., Rotterdam, vol. 1, pp. 313–21.

BLACK, R. F. (1954). "Permafrost – A Review," *Bull. Geol. Soc. Am.*, vol. 65, no. 1, Sept., pp. 839–56.

—— (1956). "Permafrost as a Natural Phenomenon," *The Dynamic North,* Tech. Asst. to Chief of Naval Operations for Polar Projects, June, Book II, Chap. I, pp. 1–25.

BATEMAN, J. D. (1949). "Permafrost at Giant Yellowknife," *Trans. Roy. Soc., Can.*, vol. 43, ser. III, sect. 4, June, pp. 7–11.

BERTON, PIERRE (1956). *The Mysterious North*. Toronto, McClelland and Stewart, 345 pp.

BONDAREV, P. D. (1959). "A General Engineering Geocryological Survey of the Permafrost Regions of the U.S.S.R. and Methods of Construction in Permafrost Areas," *Problems of the North*, no. 3, pp. 23–47.

BOSTOCK, H. S. (1948). "Physiography of the Canadian Cordillera with Special Reference to the Area North of the Fifty-Fifth Parallel," *Geol. Surv. of Can.*, Memoir 247, 106 pp.

—— (1964). "A Provisional Physiographic Map of Canada," *Geol. Surv. of Can.*, Paper 64–35, 24 pp.

BOULDING, J. D. (1963). "Here There's More to Mining than Digging Ore," *Northern Miner*, 28 Nov., pp. 21, 29.

BOYLE, R. W. (1955). "Permafrost, Oxidation Phenomena, and Hydrogeo-chemical Prospecting in the Mayo Area, Yukon," *Bull. Geol. Soc. Am.*, vol. 66, no. 12, part II, Dec., pp. 1701 (abstract).

—— (1956). "Geology and Geochemistry of Silver-Lead-Zinc Deposits of Keno Hill and Sourdough Hill, Yukon Territory," Geol. Surv. of Can., Dept. of Mines and Tech. Surveys, Ottawa, Canada, Paper 55–30, 78 pp.

BRANDON, L. V. (1965). "Groundwater Hydrology and Water Supply in the District of Mackenzie, Yukon Territory and Adjoining Parts of British Columbia," *Geol. Surv. of Can.*, Paper 64–39.

BRITTON, M. E. (1957). "Vegetation of the Arctic Tundra," *Arctic Biology*, Biol. Colloquium, ed. H. P. Hansen, Oregon State Coll., Corvallis, Ore., pp. 26–61.

BROWN, C. J. (1965). "Yukon Mining Survey – 1965," speech presented to Junior Investment Dealers Assoc. of Can. (reference unknown).

BROWN, R. A., and M. E. WOPNFORD (1955). "Operational Problems in Oil Exploration in the Northwest Territories," *Can. Mining and Metallurgical Bull.*, vol. 48, no. 524, Dec., pp. 783–9.

BROWN, R. J. E. (1956). "Permafrost Investigations in the Mackenzie Delta," *Can. Geogr.*, no. 7, pp. 21–6.

—— (1960). "The Distribution of Permafrost and Its Relation to Air Temperature in Canada and the U.S.S.R.," *Arctic*, vol. 13, no. 3, Sept., pp. 163–77.

—— (1964). "Permafrost Investigations on the Mackenzie Highway in Alberta and Mackenzie District," Div. Bldg. Res., Nat. Res. Council Tech. Paper 175 (NRC 7885).

—— (1965a). "Some Observations on the Influence of Climatic and Terrain Features on Permafrost at Norman Wells, N.W.T., Canada," *Can. J. Earth Sci.*, vol. 2, no. 1, pp. 15–31 (NRC 8213).

—— (1965b). "Permafrost Investigations in Saskatchewan and Manitoba," *Div. Bldg. Res.* Nat. Res. Council Tech. Paper 193 (NRC 8375).

—— (1965c). "Factors Influencing Discontinuous Permafrost in Canada," Symp. on Cold Climate Processes and Environments, Alaska Field Conf. (F), International Assoc. for Quaternary Res. (INQUA), Fairbanks, Alaska, Aug., 40 pp.

—— (1966a). "Influence of Vegetation on Permafrost," *Proc. Permafrost International Conf.*, Nat. Acad. of Sci., Nat. Res. Council Publ. no. 1287, Nov., pp. 20–5 (NRC 9274).

—— (1966b). "Permafrost as an Ecological Factor in the Subarctic," UNESCO Symp. on the Ecology of Subarctic Regions, Otaniemi (Helsinki) Finland, 25 July– 3 Aug., 22 pp.

—— (1966c). "Relation between Mean Annual Air and Ground Temperatures in the Permafrost Region of Canada," *Proc. Permafrost International Conf.*, Nat. Acad. of Sci., Nat. Res. Council Publ. no. 1287, pp. 241–7.

—— (1967a). "Permafrost Investigations in British Columbia and Yukon Territory," *Div. of Bldg. Res.*, Nat. Res. Council Tech. Paper 253 (NRC 9762).

—— (1967b). "Permafrost in Canada," map publ. by *Div. of Bldg. Res.*, Nat. Res. Council (NRC 9769) and *Geol. Surv. of Can.* (Map 1246A), Aug.

—— (1968a). "Occurrence of Permafrost in Canadian Peatlands," *Proc. Third International Peat Cong.*, Quebec City, P.Q., Aug. (in press).

—— (1968b). "Permafrost Investigations in Northern Ontario and Eastern Manitoba," *Div. of Bldg. Res.*, Nat. Res. Council, Tech. Paper 291 (NRC 10465).

—— (1969). "Permafrost Investigations in Labrador-Ungava," *Div. of Bldg. Res.*, Nat. Res. Council (in press).

BROWN, R. J. E. and G. H. JOHNSTON (1964). "Permafrost and Related Engineering Problems," *Endeavour*, vol. 23, no. 89, May, pp. 66–72 (NRC 7860).

BROWN, W. G., G. H. JOHNSTON, and R. J. E. BROWN (1964). "Comparison of Observed and Calculated Ground Temperatures with Permafrost Distribution under a Northern Lake," *Can. Geotech. J.*, vol. 1, no. 3, July, pp. 147–54 (NRC 8129).

CAMERON, R. C. (1969). "Cementing Well Casing in Permafrost," *Proc. Third Can. Conf. on Permafrost*, Assoc. Ctee. on Geotech. Res., Nat. Res. Council (in press).

CANADA (1909). *The Yukon Territory – Its History and Resources.* Govt. Printing Bureau, Ottawa, 181 pp.

—— (1957). *Permafrost – A Digest of Current Information.* Directorate of Engineering Development, Dept. of Nat. Defence, Ottawa, 1949. Reissued as Tech. Mem. no. 49 of the Assoc. Ctee. on Soil and Snow Mechanics, Ottawa, Aug., 48 pp.

Canadian Mining and Metallurgical Bulletin (1961). "Current Operations at United Keno Hill Mines," vol. 54, no. 594, Oct., pp. 1–18.

CHAMBERS, E. J., ed. (1907). *Canada's Fertile Northland.* Report of Evidence Heard before a Select Committee of the Senate of Canada during the Parliamentary Session of 1906–7, Published under Direction of R. K. Young, DLS, Supt. of Railway Lands Branch, Dept. of Interior, Ottawa, 139 pp.

—— (1914). *The Unexploited West.* Published under Direction of F. C. C. Lynch, Supt. of Railways Lands Branch, Dept. of Interior, Ottawa, 361 pp.

CHARLES, J. L. (1959). "Permafrost Aspects of Hudson Bay Railroad," *J. of the Soil Mechs. and Foundations Div.*, *Proc. Am. Soc. Civil Eng.*, vol. 38, no. SM 6, Dec., part 1, pp. 125–35.

COLLINS, F. H. (1955). *The Yukon Territory,* A Brief Presented to the Royal Commission on Canada's Economic Prospects, Edmonton, Alta., 22 Nov., 30 pp.

Construction World (1963). "They're Going to Beat the Bog in Manitoba Road Building," Feb., pp. 23.

COOK, F. A. (1958). "Temperatures in Permafrost at Resolute, N.W.T.," *Geogr. Bull.*, no. 12, pp. 5–18.

COOKE, H. C. and W. A. JOHNSTON. (1933). *Gold Occurrence of Canada, Summary Account* (2nd ed., 1933). Geol. Surv. of Can., Dept. of Mines, Canada, Economic Geology Series no. 10, 71 pp.

COPLAND, A. (1956). *Natural Cold Storage in the Canadian North.* Defence Res. Board, Ottawa, Rept. no. 3/55, Feb., 6 pp.

COPP, S. C., C. B. CRAWFORD, and J. W. GRAINGE (1956). "Protection of Utilities against Permafrost in Northern Canada," *J. Am. Water Works Assoc.*, vol. 48, no. 9, pp. 1155–88.

DAWSON, C. A., ed. (1947). *The New North-West*. University of Toronto Press, Toronto, 341 pp.

DAY, J. H. (1963). "Pedogenic Studies on Soils Containing Permafrost in the Mackenzie River Basin," *Proc. First Can. Conf. on Permafrost*, Assoc. Ctee. on Soil and Snow Mechs., Nat. Res. Council, Tech. Memo. 76, Jan., pp. 37–42.

DAY, J. H. and A. LEAHEY. (1957). *Reconnaissance Soil Survey of the Slave River Lowland in the Northwest Territories of Canada*. Experimental Farms Service, Dept. of Agriculture, Ottawa, Can., 44 pp.

DAY, J. H. and H. M. RICE. (1964). "The Characteristics of Some Permafrost Soils in the Mackenzie Valley, N.W.T.", *Arctic*, vol. 17, no. 4, pp. 222–36.

DENNY, C. S. (1952). "Late Quaternary Geology and Frost Phenomena Along Alaska Highway, Northern B.C. and Southeastern Y.T.," *Bull. Geol. Soc. Am.*, vol. 63, Sept., pp. 883–921.

DICKENS, H. B. (1959). "Water Supply and Sewage Disposal in Permafrost Areas of Northern Canada," *Polar Record*, vol. 9, no. 62, May, pp. 421–32.

DICKENS, H. B. and D. M. GRAY (1960). "Experience with a Pier-Supported Building over Permafrost," *J. of the Soil Mechs. and Foundation Div., Proc. Am. Soc. Civil Eng.*, vol. 86, no. SM 5, Oct., pp. 1–14.

DIER, J. S. (1969). "Techniques for Setting Drilling Rig Piling and Surface Casing under Permafrost Conditions," *Proc. Third Can. Permafrost Conf.*, Assoc. Ctee. on Geotech. Res., Nat. Res. Council (in press).

DOSTOVALOV, B. N. and A. I. POPOV (1966). "Polygonal Systems of Ice Wedges and Conditions of their Development," *Proc. Permafrost International Conf.*, Nat. Acad. of Sci., Nat. Res. Council Publ. no. 1287, pp. 102–5.

DREWE, J. G. (1969). "Design and Construction Problems at the Clinton Mine, Cassiar Asbestos Corporation Limited," *Proc. Third Can. Conf. on Permafrost*, Assoc. Ctee. on Geotech. Res., Nat. Res. Council (in press).

DUBNIE, A. and W. K. BUCK (1965). "Progress of Mineral Development in Northern Canada," *Polar Record*, vol. 12, no. 87, pp. 683–702.

EAGER, W. L. and W. T. PRYOR (1965). "Ice Formation on the Alaska Highway," *Public Roads*, vol. 24, no. 3, Jan.–Mar., pp. 55–74, 82.

ECKENFELDER, G. V. (1950). "Snare River Power Project," *Eng. J.*, vol. 33, no. 2, Feb., pp. 165–71.

Engineering News Record (1962). "Aluminum Pays in Wilderness Pipe," 3 May, pp. 35–6.

ESPLEY, G. H. (1969). "Experience with Permafrost in Gold Mining," *Proc. Third Can. Conf. on Permafrost*, Assoc. Ctee. on Geotech. Res., Nat. Res. Council (in press).

FAGIN, K. M. (1947). "Drilling Problems in Alaska," *Petroleum Eng.*, vol. 19, no. 1, Oct., pp. 180, 182, 184, 186, 188, 190.

FERRIANS, O. J., JR. (1965). "Permafrost Map of Alaska," US Geol. Surv., Misc. Geol. Investigations Map I–445.

FLEMING, H. A. (1957). *Canada's Arctic Outlet*. Berkeley, University of California Press, 129 pp.

FLINT, R. F. (1957). "Glacial and Pleistocene Geology," New York, John Wiley and Sons, 553 pp.

FROST, R. E. (1952). "Interpretation of Permafrost Features from Air Photographs," *Frost Action in Soils*, US NAS-NRC Highway Res. Board, Spec. Rept. no. 2, pp. 223–46.

—— (1960). "Aerial Photography in Arctic and Subarctic Engineering," *J. Air Transport Div.*, Am. Soc. Civil Eng., vol. 86, pp. 27–56.

FYLES, J. G. (1963). "Surficial Geology of Victoria and Stefanson Islands, District of Franklin," Geol. Surv. of Can., Dept. of Mines and Tech. Surveys, Ottawa, Canada, Bull. 101, 38 pp.

GASSER, G. W. (1948). "Agriculture in Alaska," *Arctic*, vol. 1, no. 1, pp. 75–83.

GRAINGE, J. W. (1959). "Water Supplies in the Central and Western Canadian North," *J. Am. Water Works Assoc.*, vol. 51, no. 1, Jan., pp. 55–66.

GUIMOND, R. (1958a). "Cyrus Eaton Operations are Most Northerly in Quebec-Labrador Trough," *Precambrian*, vol. 31, no. 10, Oct. pp. 32–4, 36, 38, 43.

—— (1958b). "Iron Ore Company Project Opens Vast Part of Quebec-Labrador Trough," *Precambrian*, vol. 31, no. 10, Oct., pp. 44–6, 48–52, 54–6.

GUTSELL, B. (1953). "Dawson City," *Geogr. Bull.*, no. 3, pp. 23–35.

HANNAH, G. J. R. (1961). "The New North: Land of Today – North Rankin Nickel Mines Limited," *Precambrian, Mining in Canada*, Dec.–Jan., pp. 6–9, 11, 14–7, 20–1.

HANSEN, H. P. (1953). "Postglacial Forests in the Yukon Territory and Alaska," *Am. J. Sci.*, vol. 251, pp. 505–42.

HARDING, R. (1963). "Foundation Problems at Fort McPherson, N.W.T.," *Proc. First Can. Conf. on Permafrost*, Assoc. Ctee. on Soil and Snow Mechanics, Nat. Res. Council, Tech. Memo. 76, Jan., pp. 159–66.

HARE, F. K. (1952). "The Labrador Frontier," *Geogr. Rev.*, vol. 42, no. 3, pp. 405–24.

HARWOOD, T. A. (1955). *Geology and Physiography of the Arctic Region of North Continental America and Greenland.* Arctic Sect., Defence Res. Board, Dept. of Nat. Defence, Can., Rept. no. 1/55, Nov., 31 pp.

—— (1969). "Some Possible Problems with Pipelines in Permafrost Regions," *Proc. Third Can. Conf. on Permafrost*, Assoc. Ctee. on Geotech. Res., Nat. Res. Council (in press).

HATHERTON, T. (1960). "Electrical Restivity of Frozen Earth," *J. of Geophysical Res.*, vol. 65, no. 9, Sept., pp. 3023–4.

Heavy Construction News (1964). "Plan First Permafrost Dam," 5 June, pp. 21, 24.

HEMSTOCK, R. A. (1952). "Permafrost Problems in Oil Development in Northern Canada," *Can. Mining and Metallurgical Bull.*, vol. 45, no. 481, May, pp. 280–3.

—— (1953). *Permafrost at Norman Wells,* N.W.T., Imperial Oil Limited, Calgary, 100 pp. (prepared originally in Feb., 1949).

HOBSON, G. D. (1962). "Seismic Exploration in the Canadian Arctic Islands," *Geophys.*, vol. 27, no. 2, pp. 253–73.

HOPPER, H. R. (1961). "Kelsey: Power for Northern Manitoba," *Can. Geogr. J.*, vol. 63, no. 5, Nov., pp. 170–81.

HUMPHREYS, G. (1958). "Schefferville, Quebec: A New Pioneering Town," *Geogr. Rev.*, vol. 48, no. 2, Apr., pp. 151–66.

HUSTICH, I. (1953). "The Boreal Limits of Conifers," *Arctic*, vol. 6, pp. 149–62.

HUTTON, V. F. (1946). *Report on Agricultural Investigations in the Mackenzie River Basin 1945.* Central Experimental Farm, Dept. of Agriculture, Ottawa, Canada, Feb., 17 pp. (mimeo.).

HVORSLEV, M. J. and T. B. GOODE (1966). "Core Drilling in Frozen Soils," *Proc.*

Permafrost International Conf., Nat. Acad. of Sci., Nat. Res. Council Publ. no. 1287, pp. 364–71.

HYLAND, W. L. and M. H. MELLISH (1949). "Steam-Heated Conduits – Utilidors – Protect Service Pipes from Freezing," *Civil Eng.*, Jan., pp. 27–9, 73.

INNIS, H. A. (1930). "Economics of the Hudson Bay Railway," *Can. Eng.*, vol. 59, no. 22, 25 Nov., pp. 659–60.

—— (1936). "Settlement and the Mining Frontier," in W. A. Mackintosh and W. L. G. Joerg, eds., *Canadian Frontiers of Settlement*, vol. 9, part II, Toronto, Macmillan, pp. 171–412.

ISHAM, J. (1949). "James Isham's Observations on Hudson's Bay, 1743," and "Notes and Observations on a Book Entitled a Voyage to Hudson's Bay in the Dobbs Galley, 1749," ed. by E. E. Rich and A. M. Johnson, Champlain Soc. for Hudson's Bay Record Soc.

IVES, J. D. (1962). "Iron Mining in Permafrost, Central Labrador–Ungava," *Geogr. Bull.*, no. 17, pp. 66–77.

JACOBSEN, G. (1963). "Deep Permafrost Measurement in North America," *Polar Record*, vol. 11, no. 74, pp. 595–96.

JOESTING, H. R. (1954). "Geophysical Exploration in Alaska," *Arctic*, vol. 7, nos. 3–4, pp. 165–75.

JOHNSTON, A. V. (1964). "Some Economic and Engineering Aspects of the Construction of New Railway Lines in Northern Canada, with Particular Reference to the Great Slave Lake Railway," *Proc. Inst. Civil Eng.*, vol. 29, Nov., pp. 571–88. Discussion by R. J. E. Brown, *Proc. Inst. Civil Eng.*, vol. 32, Sept. 1965, pp. 141–3.

JOHNSTON, G. H. (1963*a*). "Instructions for the Fabrication of Thermocouple Cables for Measuring Ground Temperatures," Div. of Bldg. Res., Nat. Res. Council Tech. Paper 157, Sept., 11 pp. (NRC 7561).

—— (1963*b*). "Soil Sampling in Permafrost Areas," Div. of Bldg. Res., Nat. Res. Council Tech. Paper no. 155, Nat. Res. Council, July, 4 pp.

—— (1965). "Permafrost Studies at the Kelsey Hydro-electric Generating Station – Research and Instrumentation," Div. of Bldg. Res., Nat. Res. Council Can., Tech. Paper 178 (NRC 7943).

—— (1966*a*). "Compact Ground Temperature Recorder," *Can. Geotech. J.*, vol. 3, no. 4, Nov., pp. 246–50 (NRC 9316).

—— (1966*b*). "Engineering Site Investigations in Permafrost Areas," *Proc. Permafrost International Conf.*, Nat. Acad. of Sci., Nat. Res. Council, Publ. no. 1287, pp. 371–4 (NRC 9270).

—— (1966*c*). "Pile Construction in Permafrost," *Proc. Permafrost International Conf.*, Nat. Acad. of Sci., Nat. Res. Council Publ. no. 1287, pp. 477–80 (NRC 9269).

—— (1969). "Dykes on Permafrost-Kelsey Generating Station," *Can. Geotech. J.*, vol. 6, no. 2, May, pp. 139–57.

JOHNSTON, G. H. and R. J. E. BROWN (1964). "Some Observations on Permafrost Distribution at a Lake in the Mackenzie Delta, N.W.T., Canada," *Arctic*, vol. 17, no. 3, pp. 163–75.

—— (1965). "Stratigraphy of the Mackenzie River Delta, Northwest Territories, Canada," *Geol. Soc. Am. Bull.*, vol. 76, no. 1, Jan., pp. 103–12 (NRC 8280).

—— (1966). "Occurrence of Permafrost at an Arctic Lake," *Nature*, vol. 211, no. 5052, 27 Aug., pp. 952–3.

JOHNSTON, G. H., R. J. E. BROWN, and D. N. PICKERSGILL (1963). "Permafrost Investigations at Thompson, Manitoba: Terrain Studies," Div. Bldg. Res., Nat. Res. Council, Tech. Paper 158 (NRC 7568).

KELLOGG, C. E. and I. J. NYGARD (1951). *Exploratory Study of the Principal Soil Groups of Alaska*. Agricultural Monograph no. 7, US Dept. of Agriculture, Mar., 138 pp.

KERSTEN, M. S. (1966). "Thermal Properties of Frozen Ground," *Proc. Permafrost International Conf.*, Nat. Acad. of Sci., Nat. Res. Council Publ. no. 1287, pp. 301–5.

KILGOUR, R. J. (1969). "Mining Experience with Permafrost," *Proc. Third Can. Conf. on Permafrost*, Assoc. Ctee. on Geotech. Res., Nat. Res. Council (in press).

KLASSEN, H. P. (1965). "Public Utilities Problems in the Discontinuous Permafrost Areas," *Proc. Can. Regional Permafrost Conf.*, Assoc. Ctee. on Soil and Snow Mechs., Nat. Res. Council, Tech. Memo. 86, Sept., pp. 106–18.

KOROL, N. (1955). "Agriculture in the Zone of Perpetual Frost," *Science*, vol. 122, no. 3172, Oct., pp. 680–2.

LACHENBRUCH, A. II. (1957). "Thermal Effects of the Ocean on Permafrost," *Bull. Geol. Soc. Am.*, vol. 68, no. 11, pp. 1515–30.

—— (1966). "Contraction Theory of Ice-Wedge Polygons: A Qualitative Discussion," *Proc. Permafrost International Conf.*, Nat. Acad. of Sci., Nat. Res. Council Publ. no. 1287, pp. 63–71.

LACHENBRUCH, A. H., M. C. BREWER, G. W. GREENE, and B. V. MARSHALL (1962). "Temperatures in Permafrost," in *Temperature – Its Measurement and Control in Science and Industry*, vol. 3, part I, pp. 791–803, New York, Reinhold Publishing Corp.

LANG, A. H. (1961). "Geology of Canada," reprinted from *Canada Year Book*, 14 pp.

LAWRENCE, R. D. and J. A. PIHLAINEN (1963). "Permafrost and Terrain Factors in a Tundra Mine Feasibility Study," *Proc. First Can. Conf. on Permafrost*, NRC Assoc. Ctee. on Soil and Snow Mechs., Nat. Res. Council Tech. Memo. 76, Jan., pp. 207–14.

LEAHEY, A. (1943). *Preliminary Report of an Exploratory Soil Survey along the Alaska Military Highway and the Yukon River System*. Experimental Farms Service, Dept. of Agriculture, Ottawa, 13 Oct., 16 pp. (mimeo.).

—— (1944). *Report of an Exploratory Survey Made along the Water Route from Fort Nelson, B.C., to Waterways, Alberta, Via Fort Simpson*. Experimental Farms Service, Dept. of Agriculture, Ottawa, 26 pp.

—— (1953). *Preliminary Soil Survey of Land Adjacent to the Mackenzie Highway in the N.W.T.* Experimental Farms Service, Dept. of Agriculture, Ottawa, Jan., 22 pp. (mimeo.).

—— (1954). *Soils of the Arctic and Subarctic Regions of Canada*. Soils and Agricultural Engineering Division, Division of Field Husbandry, Dept. of Agriculture, Ottawa, Seminar presented on 13 Jan., 17 pp. (mimeo.).

LEECHMAN, D. (1948). "Old Crow's Village," *Can. Geogr. J.*, vol. 37, no. 1, July, pp. 2–16.

LEFFINGWELL, E. DE K. (1919). *"The Canning River Region, Northern Alaska,"* US Geol. Surv. Prof. Paper 109.

LEFROY, GENERAL SIR J. H. (1889). "Report upon the Depth of Permanently Frozen Soil in the Polar Regions: Its Geographical Limits, and Relation to the Present Poles of Greatest Cold," *Proc. Roy. Geogr. Soc.*, vol. 8, pp. 740–6.

LEGGET, R. F. (1966). "Permafrost in North America," *Proc. Permafrost International Conf.*, Nat. Acad. of Sci., Nat. Res. Council Publ. no. 1287, pp. 2–7.

LEGGET, R. F. and H. B. DICKENS (1959). *Building in Northern Canada*. Tech. Paper no. 62, Div. of Bldg. Res., Nat. Res. Council, Ottawa, Mar. (first edition), 41 pp. (NRC 5108).

LEGGET, R. F. and J. A. PIHLAINEN (1960). "Some Aspects of Muskeg Research in Permafrost Studies," *Proc. Sixth Muskeg Research Conf.*, Assoc. Ctee. on Soil and Snow Mechs., NRC, 20 and 21 Apr., Calgary, Alta., 13 pp.

LEGGET, R. F., H. B. DICKENS, and R. J. E. BROWN (1961). "Permafrost Investigations in Canada," *Geol. Arctic*, vol. 11, pp. 956–69.

LINDQVIST, S. and J. O. MATTSSON (1965). "Studies on the Thermal Structure of a Pals," *Lund Studies in Geography*, ser. A, Phys. Geog. no. 34, pp. 38–49.

LINELL, K. A. (1957). "Airfields on Permafrost," *Proc. Am. Soc. Civil Eng. J.* Air Transport Div., vol. 83, no. 1, July, pp. 1326–1 to 1326–15.

LORD, C. S. (1941). *Mineral Industry of the Northwest Territories*. Geol. Surv. of Can., Dept. of Mines, Canada, Memoir 230, 136 pp.

LOVE, H. W. (1954). "The Northwest Highway System," *Eng. J.*, vol. 37, June, pp. 671–7.

LUND, J. (1951). "Cold Water Thawing of Frozen Placer Gravel," *Can. Mining and Metallurgical Bull.*, vol. 44, no. 468, Apr., pp. 273, 277.

MAIR, A. (1965). "Dam for Northwest Territories," *Eng. and Contract Record*, Mar., pp. 68–9.

MACDONALD, D. H. (1966). "Design of Kelsey Dykes," *Proc. International Permafrost Conf.*, Nat. Acad. of Sci., Nat. Res. Council Publ. no. 1287, pp. 492–6.

MACDONALD, D. H., R. A. PILLMAN, and H. R. HOPPER (1960). "Kelsey Generating Station Dam and Dykes," *Eng. J.*, vol. 43, no. 10, Oct., pp. 87–98.

MACKAY, J. R. (1962). "Pingos of the Pleistocene Mackenzie River Delta Area," *Geogr. Bull.*, no. 18, pp. 21–63.

—— (1963). "The Mackenzie Delta Area, N.W.T.," *Geogr. Br., Dept. Mines and Tech. Surv.*, Memo. no. 8, 202 pp.

MCDONALD, D. C. (1953). "Mining at Giant Yellowknife," *Mining and Metallurgical Bull.*, vol. 46, no. 492, Apr., pp. 199–209.

MERRILL, C. L., J. A. PIHLAINEN, and R. F. LEGGET (1960). "The New Aklavik–Search for the Site," *Eng. J.*, vol. 43, no. 1, Jan., pp. 52–7.

Mining in Canada. (1967a). "Cassiar Asbestos Corporation Now Has Two Mines in Western Canada," Oct., pp. 31, 33.

—— (1967b). "Iron Ore Operations–Where They Have Grown in Canada," Nov., p. 17.

—— (1967c). "Largest Yukon Mine," Sept., p. 38.

MÜLLER, F. (1959). "Observations on Pingos," *Meddelelser om Grønland*, vol. 153, no. 3, 127 pp.

MULLER, S. W. (1945). *Permafrost or Permanently Frozen Ground and Related Engineering Problems*. US Geol. Surv. Spec. Rept., Strategic Eng. Study no. 62, 2nd ed., 231 pp.

NEES, L. A. and A. JOHNSON (1951). *Preliminary Foundation Exploration in Arctic*

Regions, Symp. Surface and Subsurface Reconnaissance, ASTM, Spec. Tech. Publ. no. 122, pp. 28–39.

NEWMAN, R. (1963). "Permafrost Forms Base in Northern Road Test," *Heavy Construction News*, 8 Nov., p. 21.

NICHOLS, D. R. and L. A. YEHLE (1961). "Highway Construction and Maintenance Problems in Permafrost Regions," *Proc. 12th Annual Symp. Geol. Applied Highway Eng.*, Bull. no. 24, Oct., pp. 19–29.

NICKLE, C. O. (1961). "Canada's Far Northern Oil and Gas Future," *Oil in Canada*, vol. 13, no. 45, 14 Sept., pp. 36–40.

Northern Miner (1962). "Major Yukon Iron Ore Discovery has Mining People All Agog," 26 July, pp. 1, 12.

—— (1966). "Engineer-Prospector Earns Spurs in Ungava," 24 Nov., pp. 20, 21.

—— (1967). "Major Copper Rush in Canada's Arctic, PCE Deal Budgets for $2.75 Million," 30 Mar., pp. 1, 2.

NOWOSAD, F. S. (1959). "Farming in the Subarctic," *Agricultural Inst. Rev.*, Jan.–Feb., pp. 11–14, 53.

—— (1962). "Growing Vegetables on Permafrost," *North*, vol. 10, no. 4, July–Aug., pp. 42–6.

PATTY, E. N. (1945). "Placer Mining in the Subarctic," *Western Miner*, vol. 18, no. 3, Apr., pp. 44–9.

PETROV, L. S. and L. I. RAKITOV (1940). "Usloviya Zaleganiya Nefti i Osnovnye Voprosy Razvedki i Razrabotki Neftyanykh Mestorozhdeniy v Arktike" ("Oil Fields in the Arctic, Their Survey and Exploration"), *Problemy Arktiki* (*Problems of the Arctic*), no. 3, pp. 98–109 (text in Russian).

PÉWÉ, T. L. (1954). "Effect of Permafrost on Cultivated Fields, Fairbanks Area, Alaska." US Geol. Surv. Bull. 989–F, pp. 315–51.

—— (1966). "Ice Wedges in Alaska–Classification, Distribution, and Climatic Significance," *Proc. Permafrost International Conf.*, Nat. Acad. of Sci., Nat. Res. Council Publ. no. 1287, pp. 76–81.

PIHLAINEN, J. A. (1951). "Building Foundations on Permafrost, Mackenzie Valley, N.W.T.," Tech. Rept. no. 8, Div. of Bldg. Res., Nat. Res. Council, Ottawa, June, 37 pp. (DBR 22).

—— (1955). "Permafrost and Buildings." Better Building Bull. no. 5, Div. of Bldg. Res., Nat. Res. Council, Ottawa, Sept., 27 pp.

—— (1959). "Pile Construction in Permafrost," *Proc. of the Am. Soc. Civil Eng.*, *J. of the Soil Mechs. and Foundations Div.*, vol. 85, no. SM 1, part 1, Dec., pp. 75–95 (NRC 5515).

(1961). "Fort Simpson, N.W.T.–Engineering Site Information, Soils and Permafrost Conditions," *Div. Bldg. Res.*, Nat. Res. Council Tech. Paper 126 (NRC 6452).

—— (1962). "Inuvik, N.W.T.–Engineering Site Information," Div. Bldg. Res., Nat. Res. Council, Tech. Paper 135 (NRC 6757).

PIHLAINEN, J. A., R. J. E. BROWN, and G. H. JOHNSTON (1956). *Soils in Some Areas of the Mackenzie River Delta Region*. Tech. Paper no. 43, Div. Bldg. Res., Nat. Res. Council, Ottawa, Oct., 26 pp. (NRC 4096).

PIHLAINEN, J. A., R. J. E. BROWN, and R. F. LEGGET (1956). "Pingo in the Mackenzie Delta, N.W.T.," *Bull. Geol. Soc. Am.*, vol. 67, pp. 1119–22.

PIHLAINEN, J. A. and G. H. JOHNSTON (1954). "Permafrost Investigations at Aklavik

(Drilling and Sampling)," Div. Bldg. Res., Nat. Res. Council Tech. Paper 16 (NRC 3393).

—— (1963). *Guide to a Field Description of Permafrost*, Assoc. Ctee. on Soil and Snow Mechs., Nat. Res. Council, Tech. Memo. 79, 21 pp.

PIKE, A. E. (1966). "Mining in Permafrost", *Proc. Permafrost International Conf.*, Nat. Acad. of Sci., Nat. Res. Council Publ. no. 1287, pp. 512–5.

Polar Record (1967). "Electric Power in the Northwest Territories and Yukon Territory," vol. 13, no. 86, May, pp. 654–5.

PORKHAEV, G. V. and A. V. SADOVSKIY (1959). "Beds for Roads and Airfields," Principles of Geocryology (Permafrost Studies), part II, Engineering Geocryology, pp. 231–54, Acad. of Sci. of the USSR, Moscow (Nat. Res. Council Tech. Transl. 1220).

PRITCHARD, G. B. (1962). "Inuvik, Canada's New Arctic Town," *Polar Record*, vol. 11, no. 71, May, pp. 145–54.

—— (1966). "Foundations in Permafrost Areas," *Proc. Permafrost International Conf.*, Nat. Acad. of Sci., Nat. Res. Council Publ. no. 1287, pp. 515–8.

PRYER, R. W. J. (1966). "Mine Railroads in Labrador-Ungava," *Proc. Permafrost International Conf.*, Nat. Acad. of Sci., Nat. Res. Council Publ. no. 1287, pp. 503–8.

RAISON, A. V. (1959). "Construction over Permafrost," *Roads and Eng. Construction*, vol. 97, no. 1, Jan., pp. 27–34, 106–8.

REED, J. C. (1969). "Permafrost and Pet 4," *Proc. Third Can. Conf. on Permafrost*, Assoc. Ctee. on Geotech. Res., Nat. Res. Council (in press).

RICHARDSON, H. W. (1943). "Alcan – America's Glory Road – Part III – Construction Tactics," *Eng. News Record*, vol. 130, no. 1, Jan., pp. 131–8.

—— (1944a). "Controversial Canol," *Eng. News Record*, vol. 132, no. 20, 18 May, pp. 78–84.

—— (1944b). "Finishing the Alaska Highway," *Eng. News Record*, vol. 132, no. 4, 27 Jan., pp. 94–103.

RICHARDSON, J. (1839). "Notice of a Few Observations which it is Desirable to Make on the Frozen Soil of British North America," *J. Roy. Geogr. Soc.*, vol. 9, pp. 117–20.

Roads and Engineering Construction (1959). "Huge Road-Building Program will Boost Economy of Saskatchewan," Oct., pp. 78–81.

—— (1960). "Contractor Meets Challenge of Constructing a Runway in Canada's Far North," Nov., pp. 66–9.

Roads and Streets (1952). "Maintaining the Alaska Highway," vol. 95, no. 12, Dec., pp. 106–9, 118.

ROBERTSON, R. G. (1955a). "Aklavik–A Problem and Its Solution," *Can. Geogr. J.*, vol. 50, no. 6, June, pp. 196–205.

—— (1955b). *The Northwest Territories–Its Economic Prospects*. A Brief Presented to the Royal Commission on Canada's Economic Prospects at Edmonton, Alta., on 22 Nov., 47 pp.

ROBINSON, J. L. (1945a). "Agriculture and Forests of Yukon Territory," *Can. Geogr. J.*, vol. 32, no. 8, Aug., pp. 54–72.

—— (1945b). "Land Use Possibilities in Mackenzie District, N.W.T.," *Can. Geogr. J.*, vol. 32, no. 7, July, pp. 30–47.

ROBINSON, J. M. (1965). "Permafrost and Forestry," *Proc. Can. Regional Permafrost*

Conf., Assoc. Ctee. on Soil and Snow Mechs., Nat. Res. Council, Tech. Memo. 86, Sept., pp. 132–5.

ROBSON, J. (1752). *Account of Six Years' Residence in Hudson's Bay from 1733–36 and 1744–47*, London, J. Payne and J. Bouquet.

ROETHLISBERGER, H. (1961). "Seismic Refraction Soundings in Permafrost near Thule, Greenland," *Geol. of the Arctic*, vol. 2, pp. 970–80.

SAGER, R. C. (1956). "Aerial Analysis of Permanently Frozen Ground," *Photogrammetric Eng.*, vol. 17, pp. 123–32.

SAMSON, L. and F. TORDON (1969). "Experience with Engineering Site Investigations in Northern Quebec and Northern Baffin Island," *Proc. Third Can. Conf. on Permafrost*, Assoc. Ctee. on Geotech. Res., Nat. Res. Council (in press).

SAVAGE, J. E. (1965). "Location and Construction of Roads in the Discontinuous Permafrost Zone, Mackenzie District, Northwest Territories," *Proc. Can. Regional Permafrost Conf.*, Assoc. Ctee. on Soil and Snow Mechs., Nat. Res. Council, Tech. Memo. 86, Sept., pp. 119–31.

SEBASTYAN, G. Y. (1963). "Department of Transport Procedures for the Design of Pavement Facilities and Foundation Structures in Permafrost Subgrade Soil Areas," *Proc. First Can. Conf. on Permafrost*, Assoc. Ctee. on Soil and Snow Mechs., Nat. Res. Council, Tech. Memo. 76, Jan., pp. 167–206.

SHAMSHURA, G. YA (1959). "Influence of Snow Cover on the Thermal Regime of the Ground in the Taymyr Tundra," *Materialy Po Obshchemu Merzlotovedeniyu*, vol. 7, Interdepartmental Conf. on Permafrost, Acad. of Sci. of the USSR, Moscow, pp. 186–201 (text in Russian).

SHUMSKIY, P. A. (1964). "Principles of Structural Glaciology," translation from Russian by David Kraus, New York, Dover Publications Inc., 497 pp.

SHUMSKIY, P. A. and B. I. VTYURIN (1966). "Underground Ice," *Proc. Permafrost International Conf.*, Nat. Acad. of Sci., Nat. Res. Council Publ. no. 1287, pp. 108–13.

SHVETSOV, P. F. ed. (1959). "Osnovy Geokriologii" ("Principles of Geocryology"), Acad. of Sci. of the USSR, vol. 1, Moscow, 459 pp.

SHVETSOV, P. F. and I. V. ZAPOROZHTSEVA (1963). "The Recurrent Nature and Permafrost Engineering Significance of 2–3 Year Soil Temperature Increases in the Subarctic," *Problems of the North*, no. 7, pp. 21–48.

SJÖRS, H. (1959a). "Bogs and Fens in the Hudson Bay Lowlands," *Arctic*, vol. 12, no. 1, Mar., pp. 3–19.

—— (1959b). "Forest and Peatlands at Hawley Lake, Northern Ontario," *Contributions to Botany*, Bull. 171, Nat. Museum of Can., pp. 1–31.

—— (1961). "Surface Patterns in Boreal Peatlands," *Endeavour*, vol. 20, no. 80, Oct., pp. 217–24.

SPOFFORD, C. M. (1949). "Low Temperatures in Inaccessible Arctic Inflate Construction Costs," *Civil Eng.*, vol. 19, no. 1, Jan., pp. 24–6.

STANLEY, D. R. (1965). "Water and Sewage Problems in Discontinuous Permafrost Regions," *Proc. Can. Regional Permafrost Conf.*, Assoc. Ctee. on Soil and Snow Mechs., Nat. Res. Council, Tech. Memo. 86, Sept., pp. 93–105.

STEARNS, S. R. (1966). "Permafrost (Perennially Frozen Ground)," US Army Cold Regions Res. and Eng. Laboratory, Cold Regions Sci. and Eng., part I, Sect. A2, Aug., 77 pp.

STAGER, J. K. (1956). "Progress Report on the Analysis of the Characteristics and

Distribution of Pingos East of the Mackenzie Delta," *Can. Geogr.*, no. 7., pp. 13–20.

STURDEVANT, C. L. (1943). "U.S. Army's First Official Story of the Alaskan Highway," *Roads and Bridges*, vol. 81, no. 3, Mar., pp. 27–32, 62–8.

SUMGIN, M. I., S. P. KACHURIN, N. I. TOLSTIKHIN, and D. F. TUMEL (1940). *Obshcheye Merzlotovedenie (General Frost Studies)*, Akademiya Nauk SSSR (Acad. of Sci. of the USSR), Moscow, 338 pp. (text in Russian).

TEDROW, J. C. F. (1966). "Arctic Soils," *Proc. Permafrost International Conf.*, Nat. Acad. of Sci., Nat. Res. Council Publ. no. 1287, pp. 50–5.

TEDROW, J. C. F. and J. BROWN (1967). "Soils of Arctic Alaska in Arctic and Alpine Environments," in H. E. Wright, Jr. and W. H. Osburn, eds., Indiana University Press, pp. 283–93.

THOM, B. G. (1969). "Permafrost in the Knob Lake Iron Mining Region," *Proc. Third Can. Conf. on Permafrost*, Assoc. Ctee. on Geotech. Res., Nat. Res. Council (in press).

THOMPSON, H. A. (1966). "Air Temperatures in Northern Canada with Emphasis on Freezing and Thawing Indices," *Proc. Permafrost International Conf.*, Nat. Acad. of Sci., Nat. Res. Council Publ. no. 1287, pp. 18–36.

THOMSON, S. (1957). "Some Aspects of Muskeg as it Affects the Northwest Highway System," *Proc. Third Muskeg Res. Conf.*, Assoc. Ctee. on Soil and Snow Mechs., Nat. Res. Council, Tech. Memo. 47, July, pp. 42–9.

—— (1966). "Icings on the Alaska Highway," *Proc. Permafrost International Conf.*, Nat. Acad. of Sci., Nat. Res. Council Publ. no. 1287, pp. 526–9.

TRUEMAN, A. S. (1962). "Northern Road Problems Tackled by DPW Engineers," *The Dispatch*, Spring, pp. 1–4.

TSYPLENKIN, YE. I. (1944). "Vechnaya Merzlota i Ee Agronomicheskoe Znacheniye" ("Permafrost and Its Agricultural Significance"), Trudy Instituta Merzlotovedeniya im. V. A. Obrucheva (trans. of the V. A. Obruchev Inst. of Frost Studies), vol. 4, pp. 230–55 (text in Russian).

TYRRELL, J. B. (1917). "Frozen Muck in the Klondike District," *Trans. Roy. Soc. Can.*, vol. 2, sect. 4, May, pp. 39–46.

TYRTIKOV, A. P. (1959). "Perennially Frozen Ground and Vegetation," chap. XII, *Principles of Geocryology* (Permafrost Studies) part I, General Geocryology, pp. 399–421, Acad. of Sci. of the USSR, Moscow, Nat. Res. Council Tech. Transl. 1163.

UNITED STATES (1956). "Information on Alaska Pipeline for Magazine Prospectus." Information supplied by Office of Public Information, Dept. of Defence, Washington. Summarized in *Polar Record*, no. 52, pp. 43.

VIERECK, L. A. (1965). "Relationship of White Spruce to Lenses of Perennially Frozen Ground, Mount McKinley National Park, Alaska," *Arctic*, vol. 18, no. 4, Dec., pp. 262–7.

VONDER, K. L. (1953). "Operating Problems in Oil Exploration in the Arctic," *Petroleum Eng.*, vol. 25, no. 2, Feb., pp. B-12, B-14, B-16 to B-18.

VYALOV, S. S. (1965). "Realogicheskiye Svoystva i Nesushchaya Sposobnost' Merzlykh Gruntov" ("Rheological Properties and Bearing Capacity of Frozen Soils"), Acad. Sci. USSR (Moscow 1959) US Army Cold Regions Res. and Eng. Laboratory, translation no. 74, Sept., 219 pp.

WALLACE, J. J. (1961). "Road Construction for Transitional Permafrost Zones,"

Proc. Seventh Muskeg Res. Conf., NRC Assoc. Ctee. on Soil and Snow Mechs., Nat. Res. Council Tech. Memo. 71, Dec., pp. 155–63.

WASHBURN, A. L. (1947). "Reconnaissance Geology of Portions of Victoria Island and Adjacent Regions, Arctic Canada," *Geol. Soc. Am.*, Memo. 22.

—— (1956). "Classification of Patterned Ground and Review of Suggested Origins," *Bull. Geol. Soc. Am.*, vol. 67, no. 7, July, pp. 823–65.

WATMORE, T. G. (1969). "Thermal Erosion Problems in Pipelining," *Proc. Third Can. Conf. on Permafrost*, Assoc. Ctee. on Geotech. Res., Nat. Res. Council (in press).

WATTS, M. and D. K. MEGILL (1965). "Iron Ore on Baffin Island," *Can. Geogr. J.*, vol. 73, no. 5, pp. 156–65.

WEBER, W. W. and S. S. TEAL (1959). "A Sub-arctic Mining Operation," *Can. Mining and Metallurgical Bull.*, vol. 62, July, pp. 252–6.

WERNECKE, L. (1932). "Glaciation, Depth of Frost, and Ice Veins of Keno Hill and Vicinity, Yukon Territory," *Eng. and Mining J.*, vol. 133, Jan., pp. 38–43.

WHITE, L. G. (1963). "The Canada Tungsten Property, Flat River Area, Northwest Territories," *Can. Mining and Metallurgical Bull.*, vol. 50, no. 613, May, pp. 390–3.

WILLIAMS, G. P. (1966). "Some Micrometeorological Observations over Sphagnum Moss," *Proc. Eleventh Muskeg Res. Conf.*, Assoc. Ctee. on Geotech Res., Nat. Res. Council Tech. Memo. 87, May, pp. 82–91.

WILLIAMS, M. Y. (1949). "Churchill, Manitoba," *Can. Geogr. J.*, vol. 39, no. 3, Aug., pp. 122–33.

WOODS, K. B. and R. F. LEGGET (1960). "Transportation and Economic Potential in the Arctic," *Traffic Quarterly*, Oct., pp. 435–58 (NRC 5952).

WOODS, K. B., R. W. J. PRYER, and W. J. EDEN (1959). "Soil Engineering Problems on the Quebec North Shore and Labrador Railway," *Am. Railway Eng. Assoc. Bull.*, vol. 60, no. 549, Feb., pp. 669–88.

YATES, A. B. and D. R. STANLEY (1966). "Domestic Water Supply and Sewage Disposal in the Canadian North," *Proc. Permafrost International Conf.*, Nat. Acad. of Sci., Nat. Res. Council Publ. no. 1287, pp. 413–9.

YEFIMOV, A. I. and I. YE. DUKHIN (1968). "Some Permafrost Thicknesses in the Arctic," *Geologiya i Geofizika*, no. 7 (1966) pp. 92–7, reprinted in *Polar Record*, vol. 14, no. 88, Jan., p. 68.

Bibliography

ANDERSON, L. G., and P. R. MOYER (1955). "Blasting in Surface and Drift Operations in the Far North," *Dynamic North*, Tech. Asst. to Chief of Naval Operations for Polar Projects, June, book II, chap. IV, pp. 1–3.

ANDREWS, J. T. (1961). "Permafrost in Southern Labrador-Ungava," *Can. Geogr.*, vol. 5, no. 3, pp. 34–5.

ARCHIBALD, E. S. (1944). "Agricultural Lands in the Canadian Northwest," *Can. Geogr. J.*, vol. 29, no. 1, July, pp. 40–51.

BECKER, E. A. (1949). *Klondike '98.* Portland, Ore., Binsford and Mort, 127 pp.

BLACK, R. F. (1950). "Permafrost," in *Applied Sedimentation*, ed. Parker D. Trask, New York, John Wiley and Sons, pp. 247–75.

BOTSFORD, J. N. (1957). "The Gunnar Uranium Deposit," *Western Miner and Oil Rev.*, vol. 30, no. 3, March, pp. 37–45.

BROWN, R. J. E. (1957). "Observations on Break-Up in the Mackenzie River and its Delta in 1957," *J. of Glaciology*, vol. 3, no. 22, Oct., pp. 133–41.

—— (1962). "A review of Permafrost Investigations in Canada," *Can. Geogr.*, vol. 6, nos. 3–4, pp. 162–5 (NRC 7139).

—— (1966). "Permafrost, Climafrost, and the Muskeg H. Factor," *Proc. Eleventh Muskeg Res. Conf.*, Assoc. Ctee. on Soil and Snow Mechs., Nat. Res. Council Tech. Memo. 87, May, pp. 159–78.

—— (1967). "Comparison of Permafrost Conditions in Canada and the U.S.S.R.," *Polar Record*, vol. 13, no. 87, Sept., pp. 741–51 (NRC 9741).

BYRNE, N. W. (1957). "The Rayrock Story," *Western Miner and Oil Rev.*, vol. 30, no. 4, Apr., pp. 71–7.

CADELL, H. M. (1914). "The Klondike and Yukon Goldfield in 1913," *Scott. Geogr. Mag.*, vol. 30, pp. 337–56.

CAMSELL, C. (1947). *Canada's New Northwest – A Study of the Present and Future Development of Mackenzie District of the Northern Parts of Alberta and British Columbia.* North Pacific Planning Project, 155 pp.

CANADA (1907). *The Yukon Territory – Its History and Resources.* Issued by direction of the Hon. Frank Oliver, Minister of the Interior, Ottawa, 139 pp.

—— (1916). *The Yukon Territory – Its History and Resources.* Issued by direction of the Hon. W. J. Roche, Minister of the Interior, Ottawa, 233 pp.

—— (1926). *The Yukon Territory.* Northwest Territories and Yukon Branch, Dept. of the Interior, Ottawa, 100 pp.

Can. Mining J. (1952). "Moving a Town 165 Miles Northward," vol. 73, no. 3, pp. 77–80.

NOTE: The works in this supplementary list have been consulted but not used as references in the discussion in the text.

CEDERSTREM, D. J., P. M. JOHNSTON, S. SUBITZKY (1953). *Occurrence and Development of Ground Water in Permafrost Regions.* Geol. Surv. Circular 275, Geol. Surv., US Dept. of Interior, 30 pp.

CLARK, K. J. (1932). "Installing Water Mains in Perpetually Frozen Ground," *Contract Record and Eng. Rev.*, vol. 46, 30 Mar., pp. 350–1 (abstract).

Compressed Air Magazine (1954). "Water Supply Systems in the Frigid North," vol. 59, Oct., p. 295.

COOPER, S. C. (1953). "Difficulties Overcome in Building Far North Railway Line," *Roads and Eng. Construction*, vol. 91, Feb., pp. 101–7.

CRUMLISH, W. S. (1947). "Exploratory Well Drilling in Permafrost, Fort Churchill, Manitoba, Canada," *Eng. Res. and Develop.*, Lab. Rept. 1045.

CURRIE, B. W. *Prairie Provinces and Northwest Territories – Ice, Soil Temperatures.* Physics Dept., University of Saskatchewan, 28 pp. (mimeo.).

DALL, W. H., G. M. DAWSON, and W. OGILVIE (1898). *The Yukon Territory.* Downey and Co., 438 pp.

DE WET, J. P. (1945). "Mining in the North," *Beaver*, Outfit 276, Dec., pp. 20–5.

DENIS, P. Y. (1955). "Les Facteurs Géographiques de la Situation et du Site de Whitehorse" ("Geographic Factors of the Location and Site of Whitehorse"), *Rev. Can. de Géogr.*, vol. 9, no. 4, Oct.–Dec., pp. 161–78 (text in French).

DICKENS, H. B. (1960). "Construction in Permafrost: Obstacles of Soil and Climate," *Can. Consulting Eng.*, vol. 2, no. 1, Jan., pp. 33–7 (NRC 5534).

ECKENFELDER, G. V. and B. E. RUSSELL (1950). "Snare River Power Project," *Eng. J.*, vol. 33, no. 3, pp. 165–71.

FISHER, O. (1957). "The Soviet North and the Canadian North," *Can. Geogr. J.*, vol. 55, no. 3, Sept., pp. 108–15.

GARDNER, G. (1952). *Considérations sur la Valeur Economique de Grand-Nord Canadien* (Considerations of the Economic Value of Canada's North). Service de Documentation Economique, Ecole des Hautes Etudes Commerciales de Montréal, Etude no. 5, 113 pp. (text in French).

GRANT, N. A. (1953). "General Mining Conditions at Eldorado Beaverlodge," *Can. Mining and Metallurgical Bull.*, vol. 46, no. 496, Aug., pp. 490–3.

HARRINGTON, L. (1947). "North on the Hudson Bay Railway," *Can. Geogr. J.*, vol. 35, no. 2, Aug., pp. 54–66.

HEMSTOCK, R. A. (1953). "Some Northern Lights on Oil Development," *Petroleum Eng.*, vol. 25, no. 2, Feb., pp. A41–4.

JACOT, M. (1954). "Uranium City, Athabaska's Wilderness Capital," *Imperial Oil Rev.*, vol. 38, no. 3, Oct., pp. 2–8.

JENNESS, J. L. (1949). "Permafrost In Canada; Origin and Distribution of Permanently Frozen Ground, with Special Reference to Canada," *Arctic*, vol. 2, no. 1, May, pp. 13–27.

JOHNSTON, G. H. (1962). "Bench Marks in Permafrost Areas," *Can. Surveyor*, vol. 16, no. 1, Jan., pp. 32–41 (NRC 6731).

—— (1963). "Soil Sampling in Permafrost Areas," Paper presented at the Annual Meeting of the Eng. Inst. of Can., Quebec City, May. Div. of Bldg. Res., Tech. Paper no. 155 (NRC 7417).

JOHNSTON, W. A. (1930). "Frozen Ground in the Glaciated Parts of Northern Canada," *Trans. Roy. Soc. Can.*, sect. 4, pp. 31–40.

KINDLE, E. M. (1928). *Canada North of Fifty-Six Degrees*. Northwest Territories and Yukon Branch, Dept. of Interior, Ottawa, 88 pp.

LANDIS, C. S. (1945). "Blasting Deep Rock Cuts along the Alaskan Highway," *Explosives Eng.*, vol. 23, Mar.-Apr., pp. 55–76.

LEAHEY, A. (1948). *Characteristics of Soils Adjacent to the Mackenzie River in the Northwest Territories of Canada*. Experimental Farms Service, Dept. of Agriculture, Ottawa, 8 pp. (mimeo.).

—— (1949). "Factors Affecting the Extent of Arable Land and the Nature of the Soils in the Yukon Territory," *Proc. of the Seventh Pacific Sci. Congress*, vol. 6, pp. 16–20.

LEGGET, R. F. (1955*a*). "Permafrost near Lake Athabaska, Saskatchewan, Canada," *Bull. Geol. Soc. Am.*, vol. 66, p. 1589.

—— (1955*b*). "Permafrost Research," *Arctic*, vol. 7, nos. 3 and 4, pp. 153–8.

—— (1958). "Potentialities of the Northwest: An Engineering Assessment," "*Studia Varia*," ser. no. 3, Symp. on the Can. Northwest: Its Potentialities, Roy. Soc., pp. 6–22.

—— (1959). "Geology and Transportation," *Roads and Eng. Construction*, vol. 97, no. 2, Feb., 5 pp.

LEGGET, R. F., R. J. E. BROWN, and G. H. JOHNSTON (1966). "Alluvial Fan Formation near Aklavik, Northwest Territories, Canada," *Geol. Soc. Am. Bull.*, vol. 77, no. 1, Jan., pp. 15–29 (NRC 8978).

LINDSAY, J. D. and W. ODYNSKY (1965). "Permafrost in Organic Soils of Northern Alberta," *Can. J. Soil Sci.*, vol. 45, pp. 265–9.

MATHEWS, W. H. (1955). "Permafrost and its Occurrence in the Southern Coast Mountains of British Columbia," *Can. Alpine J.*, vol. 28, pp. 94–8.

MCFARLAND, W. H. S. (1939). "Operations of the Yukon Consolidated Gold Corporation," *Trans. Can. Inst. of Mining and Metallurgy*, vol. 42, pp. 537–49.

MISENER, A. D. (1955). "Heat Flow and Depth of Permafrost at Resolute, Cornwallis Island, N.W.T., Canada," *Trans. Am. Geophys. U.*, vol. 36, pp. 1055–60.

MONTEITH, H. D. (1955). "Problems in Construction in the Far North," *Eng. J.*, vol. 38, no. 6, June, pp. 784–7.

MOSS, E. H. (1953). "Marsh and Bog Vegetation in Northwestern Alberta," *Can. J. Botany*, vol. 31, pp. 448–70.

MÜLLER, F. (1962). "Analysis of Some Stratigraphical Observations and Radio Carbon Dates from Two Pingos in the Mackenzie Delta Area," *Arctic*, vol. 15, no. 4, pp. 279–88.

NORDALE, A. M. (1947). "Valuation of Dredging Ground in the Subarctic," *Trans. Can. Inst. of Mining and Metallurgy*, vol. 50, pp. 487–96.

NOWOSAD, F. S., and A. LEAHEY (1960). "Soils of the Arctic and Subarctic Regions of Canada," *Agricultural Inst. Rev.*, Mar.–Apr., pp. 48–50.

O'NEILL, P. H. (1954). "Arctic Gold Dredging," *Mining Eng.*, vol. 6, Nov., pp. 1068–71.

PATTY, E. N. (1951). "Solar Thawing Increases Profit from Subarctic Placer Gravels," *Mining Eng.*, vol. 190, no. 1, Jan., pp. 27–8.

PAYNE, H. M. (1913). "The Development and Problems of the Yukon," *Can. Mining Inst. Trans.*, vol. 16, pp. 228–40.

PEARCE, E. E. (1922). "Cold-Water Thawing of Frozen Gravel," *Mining Sci. Press*, vol. 124, 4 Feb., pp. 154–6.

PERRET, L. (1912). "Prospecting Frozen Gravel," *Mining Sci. Press*, vol. 104, 12 June, pp. 856–7.

PERRY, O. B. (1915). "Development of Dredging in Yukon Territory," *Eng. and Mining J.*, vol. 100, 25 Dec., pp. 1039–44.

PIHLAINEN, J. A. (1962). "An Approximation of Probable Permafrost Occurrence," *Arctic*, vol. 15, no. 2, pp. 151–4.

PIHLAINEN, J. A., R. J. E. BROWN, and G. H. JOHNSTON (1956). "Soils in Some Areas of the Mackenzie River Delta Region," Div. of Bldg. Res., Nat. Res. Council Tech. Paper 43 (NRC 4096).

REED, I. M. (1943). "How Dawson Keeps its Water Mains from Freezing," *Pacific Builder and Eng.*, vol. 49, no. 8, Aug., p. 54.

RICHARDSON, H. W. (1942a). "Alcan – America's Glory Road – Part I – Strategy and Location," *Eng. News-Rec.*, vol. 129, no. 25, 17 Dec., pp. 81–96.

—— (1942b). "Alcan – America's Glory Road – Part II – Supply, Equipment and Camps," *Eng. News-Rec.*, vol. 129, no. 27, 31 Dec., pp. 35–42.

—— (1953). "Build Railroad through North Canadian Bush," *Construction Methods and Equipment*, vol. 35, no. 10, Oct., pp. 62–4, 67–8, 70–2, 74–5, 78.

RICHARDSON, J. (1841). "On the Frozen Soil of North America," *Edinburgh New Phil. J.*, vol. 30, pp. 110–23.

RIDOUT, D. G. (1931). "Port Churchill," *Can. Geogr. J.*, vol. 3, no. 2, Aug., pp. 105–28.

Roads and Streets (1944). "Alaska Highway Flight Strips," vol. 87, no. 6, June, pp. 79 and 81.

ROBINSON, J. L. (1946). "Weather and Climate of the Northwest Territories," *Can. Geogr. J.*, vol. 32, no. 3, Mar., pp. 124–39.

ROCKIE, W. A. (1942). "Pitting on Alaska Farm Lands – A New Erosional Problem," *Geogr. Rev.*, vol. 32, pp. 128–34.

SPINDLER, W. H. (1944). "Drainage on the Alaska Highway," *Roads and Bridges*, vol. 82, no. 1, Jan., pp. 33–5, 80.

STEFANSSON, V. (1942). *Airfields and Roads over Frozen Subsoil*, personal communication to Scott Polar Res. Inst., Cambridge, 29 May.

STEPHENSON, W. (1957). "The Maddening War against Permafrost," *Maclean's Magazine*, 14 Sept., pp. 23, 90–3.

TABER, S. (1943). "Some Problems of Road Construction and Maintenance in Alaska," *Public Roads*, vol. 23, no. 9, July, Aug., Sept., pp. 247–51.

TERZAGHI, K. (1952). "Permafrost," *J. Boston Soc. of Civil Eng.*, vol. 39, Jan., pp. 1–50.

THOMAS, M. K. (1953). *Climatological Atlas of Canada*. Publ. jointly by the Div. of Bldg. Res., Nat. Res. Council and Meteorological Div., Dept. of Transport, Can., 253 pp. (NRC 3151).

THORLEY, A. (1963). "Review of a Recent Arctic Soil Investigation," *Proc. First Can. Conf. on Permafrost*, NRC Assoc. Ctee. on Soil and Snow Mechs., Nat. Res. Council Tech. Memo. 76, Jan., pp. 149–55.

TYRRELL, J. B. (1903). "A Peculiar Artesian Well in Klondike," *Eng. Mining J.*, vol. 75, no. 5, 31 Jan., p. 188.

Western Construction (1955). "Constructing 620 Miles of Trouble," vol. 30, no. 3, Mar., pp. 29, 32, 35, 38.

WILLIAMS, G. A. (1943). "Winter Maintenance Problems on the Alaska Highway," *Roads and Bridges*, vol. 81, no. 11, Nov., pp. 27–30, 58–9.

Index